Biosocial Aspects of Social Class

BIOSOCIAL SOCIETY SERIES
Series Editor: G.A. Harrison

Biosocial Aspects of Social Class

EDITED BY

C.G. Nicholas Mascie-Taylor
Department of Biological Anthropology
University of Cambridge

OXFORD NEW YORK TOKYO
OXFORD UNIVERSITY PRESS
1990

Oxford University Press, Walton Street, Oxford OX2 6DP

Oxford New York Toronto
Delhi Bombay Calcutta Madras Karachi
Petaling Jaya Singapore Hong Kong Tokyo
Nairobi Dar es Salaam Cape Town
Melbourne Auckland

and associated companies in
Berlin Ibadan

Oxford is a trade mark of Oxford University Press

Published in the United States
by Oxford University Press, New York

© *The Biosocial Society 1990*

British Library Cataloguing in Publication Data
Biosocial aspects of social class. – (Biosocial society
;series 2)
1. Social class
I. Mascie-Taylor, Nicholas
305.5
ISBN 0–19–857724–9

Library of Congress Cataloging in Publication Data
Biosocial aspects of social class / edited by Nicholas Mascie-Taylor.
p.   cm.—(Biosocial Society series : 2)
Includes index.
1. Social classes. 2. Family demography. 3. Sociobiology.
I. Mascie-Taylor, C. G. N.   II. Series: Biosocial Society series :
no. 2.
HT609 B45 1989 305.5—dc20   89–16075   CIP
ISBN 0–19–857724–9

Set by Colset Private Limited, Singapore
Printed in Great Britain by Boockcraft, Midsomer Norton, Avon

PREFACE

This book is the second in a series that the Biosocial Society of Great Britain has produced with the Oxford University Press. The aim of the Society is to examine topics and issues which have biological and social importance, and social class seemed a particularly apt topic for the Society to discuss.

All four contributors work on various aspects of social class. Dr Anthony Heath is particularly concerned with the conceptions of social class, its reliability and validity. He thus examines the subject from a sociologist's perspective. The next two contributions are primarily demographic. Dr Peter Goldblatt discusses mortality differences between social classes while Dr David Coleman concentrates primarily on fertility differentials. Finally Dr Mascie-Taylor looks at social class from a biological standpoint and asks whether there are genetic differences between classes.

Three of the four contributors were able to address the Biosocial Society in Oxford on 29 May 1987. Each talk was followed by a lively debate. The society is most grateful to Professor G. A. Harrison for the local arrangements including making the Pauling Centre available for the meeting.

Finally, grateful acknowledgement is made to the Parkes Foundation for a grant which made this book possible.

1989   C. G. N. M-T

CONTENTS

CONTRIBUTORS

Anthony Heath Nuffield College, Oxford, OX1 1NF

Peter Goldblatt SSRU, The City University, Northampton Square, London EC1V 0HB

David Coleman Department of Social and Administrative Studies, University of Oxford, Barnett House, Wellington Square, Oxford OX1 2ER

Nicholas Mascie-Taylor Department of Biological Anthropology, University of Cambridge, Downing Street, Cambridge CB2 3DZ

1

The Sociology of Social Class

ANTHONY HEATH

CONCEPTIONS OF SOCIAL CLASS

Class is a theory-laden concept or rather set of concepts. The different theoretical traditions in sociology have spawned different theories of class and stratification, and these in turn have spawned different measures. As a result, British sociology exhibits a bewildering array of class schemas. There are the two 'official' schemas: the Registrar General's (with its six classes) and the OPCS (Office of Population Censuses and Surveys) scheme of 17 socioeconomic groups (OPCS 1980). There is the Market Research scheme of six 'social grades' (A, B, C1, C2, D and E) (Monk 1978). From academic sociologists come the Hall–Jones classification (Hall and Jones 1950), the class schema of John Goldthorpe (Goldthorpe, Llewellyn, and Payne 1980), and the Marxist schemas of E.O. Wright (Wright 1979, 1985). Finally, there are also continuous scales, as opposed to discrete class schemas, such as the Cambridge scale (Stewart, Prandy, and Blackburn 1980) and the Hope–Goldthorpe scale (Goldthorpe and Hope 1974).

These differences are not merely the result of academic, commercial, or official wilfulness. They are in part the result of the different purposes which the inventors of the different schemes had in mind. The Registrar General's scheme was devised with mortality differentials in mind; the Market Research scheme was designed with markets for consumer goods in mind. John Goldthorpe's was designed for the study of social mobility. Wright's Marxist schema was designed to explore the potentials for collective political action.

To some extent, then, the different schemas are theoretical devices designed to explain or predict different substantive phenomena, and there is no reason in principle why the explanation of collective action should use the same schema as the prediction of individual shopping habits.

To make sense of this large array of class schemas, it may be helpful to divide them into two distinct groups. On the one hand there are schemes which conceive of society as a *hierarchy of occupations*, running from occupations of low 'standing' at the bottom (typically unskilled manual) to ones of high 'standing' (high level professionals). The underlying conception is usually that of a *continuum*, and the points on the continuum where the lines are drawn to distinguish separate classes are somewhat arbitrary. The distinctions between classes are ones of degree rather than of kind. The Registrar General's, the Market Research social grade scheme, and the Hall–Jones classification of social status are all of this type (as of course are the continuous scales).

The second group of schemes conceive of society as composed of *collectivities* with distinct and sometimes opposing *interests*. They do not see society as a hierarchy or continuum of occupations. The distinctions between classes are more of kind than degree and have to do with *conditions of employment* as well as occupation. Classes are seen as more or less *discrete* social formations, and where one draws the boundaries between classes is an empirical matter, not an arbitrary one. The main examples of this kind of scheme are the Goldthorpe and Wright classifications.

I shall not attempt to compare all these class schemas but will instead concentrate on one from each of these two groups. I shall take the Registrar General's scheme as a representative of the hierarchical classifications of occupations according to their social standing. And I shall take John Goldthorpe's as a representative of the discrete classifications of employment situations according to economic interests. (For detailed discussions of these and other schemas see Marsh 1986, Marshall 1988, and Marshall *et al*. 1988.)

The Registrar General's scheme

This scheme distinguishes six classes, as follows:

    I professional and similar occupations;
    II intermediate occupations;
    III skilled occupations;
        (NM) non-manual;
        (M) manual;
    IV partly skilled occupations;
    V unskilled occupations.

The scheme was introduced by Stevenson, Statistical Officer at the General Register Office, in 1913. (See Leete and Fox 1977; Szreter 1984 for more detailed discussions.) The problem with which Stevenson was grappling was 'the comparison of the vital statistics of the more and less prosperous sections of the same community' (Stevenson 1928, p. 207). He rejected methods of classification based on local areas, on size of tenements, on direct or indirect measures of wealth (such as numbers of servants), or on industry. He turned instead to occupation, and he grouped occupations into five classes or social grades (class III was later divided). In so far as there was any theoretical concept lying behind his classification, it appeared to be that of 'social position' or, as it was later termed, 'social standing within the community'. In turn, it appeared that 'culture or the lack of it' was the principal criterion for judging social position.

In introducing his scheme Stevenson wrote:

To some extent comparisons of vital statistics, such as infant mortality, have recently been compiled in the USA, and possibly elsewhere, with distinction of family income. . . . So far as this method can be applied it is, of course, ideal for estimation of the effects of wealth as such, apart from its cultural associations, which are much closer in some instances than in others. But its drawback is that it may fail altogether as an index to culture, probably the more important influence. The power of culture to exert a favourable influence upon mortality, even in the complete absence of wealth, is well illustrated by the case of the clergy. The income test, if it could be applied, would certainly place them well down the list, yet their mortality is remarkably low. . . . Such a record as this, consistently repeated in each succeeding report, seems to make it quite clear that the lower mortality of the wealthier classes depends less upon wealth itself than upon the culture, extending to matters of hygiene, generally on the whole associated with it. . . . But culture is more easily estimated, as between occupations, than wealth, so the occupational basis of social grading has a wholesome tendency to emphasize it. One does not hesitate to allocate the clergy, despite their unfortunately too frequent poverty, to the highest social class, and similarly in other cases regard can be paid not only to probable income but to cultured intelligence and education. It follows that when one speaks of the more or less comfortable classes one is thinking largely of the more or less cultured classes. (Stevenson 1928, pp. 208–9.)

In his reasoning, then, Stevenson seems to have moved from a concern with 'the more or less prosperous sections of the same community' to a classification of 'the more or less cultured classes'. And his reason is that culture is the more important determinant of vital statistics.

OPCS have subsequently presented somewhat different rationales for the scheme. In 1970 it was suggested that 'standing within the community' was the basic criterion:

Since the 1911 Census it has been customary, for certain analytical purposes, to arrange the large number of unit groups of the Occupational Classification into a small number of broad categories called Social Classes . . . the unit groups included in each of these categories have been selected so as to ensure that, so far as is possible, each category is homogeneous in relation to the basic criterion of the general standing within the community of the occupations concerned. This criterion is naturally correlated with, and its application conditioned by, other factors such as education and economic environment, but it has no direct relationship to the average level of remuneration of particular occupations. (OPCS 1970, p. x.)

In 1980 however, the new criterion of 'occupational skill' was introduced:

The occupation groups included in each of these categories have been selected in such a way as to bring together, so far as is possible, people with similar levels of occupational skill. (OPCS 1980, p. xi.)

These revisions of the basic criterion do not appear to have been associated with corresponding changes in the allocation of occupation groups to classes (see Brewer 1986).

John Goldthorpe's class schema

Goldthorpe's seven classes are as follows:

1 service class, higher grade;
2 service class, lower grade;
3 routine non-manual;
4 petty bourgeoisie;
5 foremen and technicians;
6 working class, skilled manual;
7 working class, semi and unskilled manual.

There are some similarities with the Registrar General (RG) scheme: for example, professionals and managers occur in the two top classes in both schemes, while semi- and unskilled manual workers occur at the bottom in both. But there are also major differences in the arrangement of the classes and in the theoretical rationale. Unlike Stevenson, Goldthorpe was not concerned with the vital statistics of birth and death but with social mobility, class formation, and class action (for example, voting in general elections). And whereas Stevenson's theory was that culture or the lack of it was the major determinant of fertility, Goldthorpe's theory was that 'market and work situations' were the primary sources of class action. He wrote:

A distinctive feature of these categories [from which the schema is constructed] is that they provide a relatively high degree of differentiation in terms of both occupational function *and* employment status: in effect, the associated employment status is treated as part of the definition of an occupation. Thus, for example, 'self-employed plumber' is a different occupation from 'foreman plumber' as from 'rank-and-file employee plumber'. On this basis, then, we are able to bring together, within the classes we distinguish, occupations whose incumbents will typically share in broadly similar *market* and *work* situations which, following Lockwood's well-known discussion, we take as the two major components of class position. That is to say, we combine occupational categories whose members would appear, in the light of the available evidence, to be typically comparable, on the one hand, in terms of their sources and levels of income, their degree of economic security and chances of economic advancement; and, on the other, in their location within the systems of authority and control governing the process of production in which they are engaged, and hence in their degree of autonomy in performing their work-tasks and roles. (Goldthorpe, Llewellyn, and Payne 1980, p. 39.)

The crucial notion of market situation is taken from Max Weber. It can perhaps be most easily explained by reference to the classes themselves. Thus classes 1 and 2 (the service class: 'the class of those exercising power and expertise on behalf of corporate bodies') are largely composed of salaried employees in secure employment within the bureaucracies of modern corporations, both public and private. These employees will usually be on incremental salary scales and enrolled in company pension schemes. (Confusingly, however, the service class also contains large employers and self-employed professionals.)

Class 3 (which is similar to the RG class IIINM) is a kind of ' "white-collar labour force", functionally associated with, but marginal to, the service class'. Its members have some of the advantages of 'staff status' associated with belonging to a bureaucracy, but they are in subordinate positions and are subject to, rather than themselves exercising, authority.

Class 4 is a complete departure from the RG scheme, and is perhaps the most interesting innovation of Goldthorpe's schema. It contains small employers and own account workers. What distinguishes this class is not their occupations as such but their conditions of employment: its members are all 'independents' directly exposed to market forces rather than cushioned by bureaucracy or trade union membership.

Classes 6 and 7 constitute the working class, rank and file employees in industry and agriculture. The basic feature of their market situation is that they are 'wage labourers'. They sell their labour to an employer in return for wages and are directly subject to the employer's authority. Class 5 on the other hand, although in many ways allied to the working class (and largely recruited from it), contains workers who have a somewhat higher degree of security and have a greater measure either of autonomy or authority in the workplace.

Thus Goldthorpe's scheme is economically-based, but occupations are allocated to classes not only on the basis of their typical income but also on the basis of the sources and security of that income and of the prospects for future advancement to more secure and economically advantaged occupations. It should be emphasized too that Goldthorpe does not see his scheme as a neatly hierarchical one. There is no straightforward sense in which class 3 ranks above class 4 or class 4 above class 5; they have *different* market and work situations but it would be wrong to think of class 3 as generally more advantaged (and certainly not in terms of income) than classes 4 or 5. On the other hand, it is true that classes 1 and 2 are more advantaged in most respects than the other classes, while class 7 is less advantaged. There is therefore a hierarchical element in the scheme, but it should not be regarded as a simple one-dimensional scheme.

It should also be noted that a close approximation to Goldthorpe's schema can be achieved using the second of the 'official' classifications, namely that of the 17 socio-economic groups. These socio-economic groups (SEGs) are differentiated according to employment status as well as occupation and can be collapsed to yield Goldthorpe's seven classes (or his eleven classes). Thus there are separate SEGs for employees and employers, for managers and foremen and so on. (SEGs 1.1, 1.2, 3, and 4 can be equated with class 1; SEGs 2.2, 5.1, and 5.2 with class 2; SEG 6 with class 3; SEGs 2.1, 12, 13, and 14 with class 4; SEG 8 with class 5; SEG 9 with class 6 and SEGs 7, 10, and 15 with class 7.)

RELIABILITY AND VALIDITY

There are various technical issues that must be considered before using any class schema, most notably those of reliability and validity. By reliability we refer to the ability of investigators to replicate each other's measurements using standardized procedures. By validity we refer to the degree to which the measurements correspond to the theoretical construct.

A precondition of reliability is that there should exist formal procedures for collecting and classifying the data. This condition is met for both the official and the Goldthorpe schemes (although it is less apparent in some other cases such as the Market Research scheme, the Hall–Jones scheme, and the Wright scheme; see Marsh 1986). Goldthorpe follows the official OPCS practice in data collection and in the coding of the basic 'atoms' of the class schemas (some 351 occupational groups). His practice differs only in the derivation of his seven classes from these basic atoms, and for this there is a routine computer program.

A number of experiments have been carried out by OPCS on the reliability of data collection and coding. In one of the most illuminating experiments two different interviewers called on the respondent, one of whom carried out interviewer coding, while the other collected the data and then sent the work to be coded in the office. When aggregated into the 17 socio-economic groups, it was found that the two measures agreed with each other in 84 per cent of cases. Of the 16 per cent discrepancy, it was found that 3 per cent related to cases where the respondent had given different job descriptions of the same job on the two occasions; 3 per cent related to ambiguous jobs, where identical information was given by the respondent but the job did not fall uniquely into one category; 1 per cent was due to inadequate probing by one or other interviewer; and 9 per cent was due to coding error (Dodd 1986).

While reliability of 84 per cent is probably as good as it is reasonable to expect, it is unlikely that the errors are altogether random. Unskilled manual jobs seem to be a particular source of error, and the employment status of foremen also appears to cause particular problems (Dodd 1986; Britton and Birch 1985; Rose *et al.* 1987).

These two sources of error are worrying, as they relate to rather small classes. Thus in Goldthorpe's scheme, foremen are the major component of his small class 5 (foremen and lower grade technicians). There is a serious risk that results for this particular class will not be reliable. Similarly, various forms of unskilled manual work are the sole constituents of the Registrar General's class V. Coding unreliability therefore makes this class, which is a particularly small one, of little value in sociological enquiry. The errors that OPCS found were between unskilled manual work and personal service work, which is allocated to class IV. RG classes IV and V should therefore be combined in any analysis. The not infrequent practice of using class V of the RG scheme as the baseline in analyses of class inequality is to be deplored.

We turn now to validity. The validation of the class schemas has usually rested on their predictive power. While predictive power is a necessary condition for a concept to be scientifically useful, it is not self-evident that it is a sufficient condition. There is the danger of circularity so that class comes to be defined simply as 'whatever best predicts differences in mortality'. This was a problem that greatly exercised the discussants of Stevenson's paper on social class (see the *Journal of the Royal Statistical Society* **91** (1928), 221–30).

In addition to predictive power, therefore, we may wish to know whether the RG scheme, for example, is indeed an ordinal ranking according to 'culture' and 'prosperity', 'standing in the community', 'skill level', or whatever it is that the current editors of the *Classification of occupations* believe their class schema to measure. Unfortunately, the editors have been rather reluctant to carry out the kind of investigation required, and one is given the impression that occupations are allocated to classes on the basis of the editors' intuitions rather than empirical evidence.

Table 1.1, however, suggests that the RG scheme is indeed an ordinal ranking with respect to 'culture', at least if culture is measured by educational qualifications, but that it is not ordinal with respect to income.[1] (See also Bland 1979.)

TABLE 1.1. The Registrar General's schema.

Respondents' class	% with 'O'-level qualifications		% earning more than £8000 p.a.	
I	74	(150)	90	(86)
II	61	(818)	64	(392)
III (NM)	51	(947)	23	(394)
III (M)	22	(801)	41	(370)
IV	17	(716)	17	(261)
V	12	(203)	8	(78)
N	3635		1583	

Notes: Figures in brackets give the *N*s for each class. Respondents are classified according to their own occupations.
Sources: Qualifications — 1987 British General Election Study[1]; income — 1987 British Social Attitudes Survey (see Jowell, Witherspoon, and Brook 1988).

We can see that there is a clear ordinal relationship between RG class and educational qualifications, although there is a marked difference between classes III(NM) and III(M). The manual non-manual distinction seems to represent a discontinuity, at least with respect to this particular measure of culture.

With respect to income, however, the picture is rather different. We can see that class III(NM) is 'out of order', with its members earning substantially less than members of class III(M). The reasons for this are not hard to find: class III(NM) contains disproportionate numbers both of young people and of women. If we control for age and sex, the anomaly is removed.

In addition to their overall ranking, one might also wish to know whether the classes are internally *homogeneous* with respect to culture or standing in the community. Are there for example particular occupational groups which are 'misplaced'? It appears that there are indeed some notable misplacements. For example, both petrol pump attendants and policemen are included in RG class III(NM) and yet one might have supposed that they were very different in their 'standing in the community', prosperity, skill, or whatever criterion one wishes to employ.

There are also cases where the basic atoms of the schema, the occupational groups, include rather heterogeneous collections of respondents. For example, a nursing sister or charge nurse is included in the same occupational group as an unqualified (and poorly paid) nursing auxiliary. (This particular anomaly may well be corrected in the next revision of the *Classification of occupations*.) Anomalies of this kind have led some critics to suggest that the classification is more appropriate for men's jobs than for women's.

The periodic revisions of the classification also raise problems of comparability over time. Comparisons over time are one of the major uses for the RG scheme and the fact that so many earlier studies used the scheme is one of the main justifications for continuing to use the scheme today. Not only, however, have the definitions of some of the atoms been changed from time to time, but also some atoms have been promoted or demoted. In 1931, for example, clerks were demoted from class II to class III; in 1961 postmen and telephone operators were demoted from class II to class IV. These changes may well be appropriate ones to take account of the changing pay and qualifications of particular occupations. As literacy has become more nearly universal, the supply of labour for jobs whose main requirement is literacy becomes much larger and the price of that labour will accordingly fall. The worry is that there may be other cases where such promotions and demotions are called for, to take account of the changing nature of the labour market. Unfortunately, OPCS has never carried out systematic assessments of the typical pay and qualifications associated with each occupation. The revisions that are made from time to time appear to be rather *ad hoc* and based on intuition rather

TABLE 1.2. Goldthorpe's class schema.

Respondents' class	Male full-time employees		
	Career prospects (% affirmative)	Autonomy (mean score)	Wage (£ mean)
1	80	5.1	14 130
2	75	4.4	10 956
3	60	3.5	7238
4	—	—	—
5	72	3.0	8469
6	36	1.9	7022
7	26	1.9	6485

Source: Marshall *et al*. (1988), Table 4.9.

than on evidence. It is not clear, therefore, that the RG scheme does provide a valid yardstick for measuring change in class differentials over time.

Some of these problems apply to Goldthorpe's schema too, since it is based on the same atoms and their revisions. However, the general face validity of his scheme can be assessed through data such as that of Table 1.2. These data come from a national survey conducted by Marshall *et al.* (1988) in which respondents were asked about various aspects of their market and work situations.

Unfortunately, Marshall and his colleagues did not ask their respondents in the petty bourgeoisie about their market and work situations. For the other classes, however, the data generally appear to support Goldthorpe's account of his classes. Thus classes 1 and 2 are generally advantaged as regards wages, career chances, and autonomy while class 7 is generally disadvantaged. Men in class 3, while earning little more than men in class 6 and being almost equally subordinate in their work situation, have much better career chances and greater autonomy. And men in class 5 clearly constitute a kind of blue-collar élite, having relatively high wages although somewhat less autonomy than class 3.

FURTHER TECHNICAL ISSUES

Before moving on to the predictive power of the different classifications, there are some further procedural matters that need to be considered. These apply to both the RG and the Goldthorpe schemas.

First, it is customary to classify people according to their current job, if they are employed, or to their last main job if retired, unemployed, or economically inactive. Secondly, it is customary to take the family as the unit of stratification, and to allocate respondents to classes on the basis of the occupation and employment status of the 'head of household', usually (in the case of married couples) taken to be the husband.

Both these procedures can be, and sometimes are, varied by investigators. The latter in particular has been the subject of lively debate. Both procedures were largely taken for granted at a time when unemployment rates were low, unemployment tended to be temporary, and few married women worked. It is not self-evident that either procedure is now so straightforward. The growth of long-term unemployment and the greater number of people who are dependent on state welfare benefits, some of whom may never have had a job, calls into question the first procedure. At the very least it means that there will be a number of respondents in any survey or census who cannot be allocated to a class. Rather than treating these respondents as a residual category and lumping them in with other sources of missing data (the usual procedure) it may be preferable to formulate explicit rules.

This can sensibly be done with Goldthorpe's schema. It would seem to be in the spirit of the schema to classify people according to their (or their household's) main source of income. Respondents whose main source of income derives from employment would be classified as at present, but additional classes at top and bottom would need to be added for those whose main source of income was either interest and dividends or was state benefits. It is doubtful, however, whether such a procedure would be theoretically appropriate for the RG scheme; it would in fact destroy its ordinal character since sources of income are not closely linked to culture or educational qualifications.

The practice of allocating people on the basis of the head of household's occupation and employment status has become extremely contentious (Acker 1980; Goldthorpe 1983; Britten and Heath 1983; Stanworth 1984). One alternative that has been suggested is to allocate respondents according to the dominance principle (Erikson 1984). That is to say, both partners are allocated to the class of the spouse with the 'higher' class. This would appear to be straightforward enough in the case of a hierarchical scheme such as the Registrar General's, but involves some questionable assumptions when used with a non-hierarchical schema such as Goldthorpe's.

Another alternative is simply to classify people according to their own occupation and employment status, if they are currently in employment, but according to their spouse's (either husband or wife's) if they are inactive (the procedure used by Heath, Jowell, and Curtice 1985). Another alternative is to use a joint classification (Britten and Heath 1983). Yet another possibility is to abandon the family as the unit of stratification and to deal solely with individuals (Stanworth 1984).

TABLE 1.3.   Alternative procedures.

| | % of respondents with an 'O'-level pass classified according to: | | | | | |
	Individual procedure		Head of household procedure		Dominance procedure	
I	74	(150)	67	(213)	68	(227)
II	61	(818)	55	(813)	56	(1051)
III (NM)	51	(947)	55	(371)	45	(872)
III (M)	22	(801)	25	(1061)	20	(834)
IV	17	(716)	22	(464)	17	(528)
V	12	(203)	20	(111)	13	(136)

Note: Figures in brackets give the *N*s for each class.
Source: British General Election Study 1987.[1]

No consensus has yet been reached on this problem. It should be noted, first, that there is substantial class homogamy between husbands and wives; secondly, that different procedures will yield different 'maps' of the class structure since the occupational distributions of men and women are rather different. Table 1.3 indicates the kinds of differences that result when the individual, dominance, and head of household procedures respectively are used.

EXPLANATORY POWER

In using any schema a crucial, perhaps *the* crucial consideration, is whether it discriminates between people in interesting respects. The popularity of social class as a concept is largely a result of the fact that it does discriminate in a wide range of interesting respects, from fertility to mortality, from educational attainment to political behaviour (see Reid 1981). To be sure, discriminatory power is not the same as explanatory power. Typically, social scientists have cross-sectional data which show a powerful association (powerful by social science standards) between measured social class and, say, ill-health. The difficulty here is that causation may run in either direction: low social class may lead to poor health or people who suffer from ill-health may be downwardly mobile. Class position may be the consequence of health, rather than the other way round.

It is impossible to resolve such issues without access to longitudinal data, and since longitudinal data also tend to suffer from serious problems of attrition, it is probably wise to be cautious even where they exist. However, it would certainly be unwise to ignore the possibility of selective mobility.

Nevertheless, a major attraction of social class is that in many cases a causal interpretation is not implausible, and is often hard to resist. This is particularly true when a temporal ordering can be assigned. Thus one major concern for sociologists has been the relation between educational attainment and social class background, as measured by the class occupied by the respondent's father before the respondent took the examinations in question. Even here, measures of the father's class are based on retrospective reports, and could in principle be subject to selective recall and there may also be selective survival (whether through death or emigration) of people from different class backgrounds.

To be on the safe side, therefore, we shall simply report patterns of association between social class and some variables of interest, and leave it to the reader to supply causal interpretations.

It is beyond the scope of this chapter to provide a full account of the relation between social class and other variables (but see Reid 1981). Instead, we shall focus on three variables only. First, we shall look at the respondents' self-assigned class to see what relation the R G and Goldthorpe classifications have to respondents' self-definitions. Secondly we shall look at educational attainment and relate this to father's social class. Since 'culture or the lack of it' appears to be one of the key concepts lying behind the R G scheme, and since the family's culture might be expected to have a major influence on children's education, this is a variable which might be expected to have a particularly close relation with the R G scheme. Thirdly, we shall look at voting behaviour. Here we might expect to find that voting behaviour has a particularly close relation with Goldthorpe's scheme since voting can be considered a form of collective action in pursuit of economic (and other) interests. For the analyses that follow our, data are taken from the 1987 British Election Survey.[1]

Table 1.4 shows how the two class schemas relate to respondents' subjective assessments of their social class positions. Respondents were asked:

Do you ever think of yourself as belonging to any particular class? [If yes] Which class is that? [If no, etc.] Most people say they belong either to the middle class or the working class. If you *had* to make a choice, would you call yourself middle class or working class?

Around half the respondents gave a middle or working class self-description spontaneously. The great majority of the remainder assented to a middle- or working-class label if they had to make a choice. In Table 1.4 the answers to the 'spontaneous' and the 'forced' parts of the question have been combined.

As we can see, there is a general tendency for respondents to prefer a working-class self-description, with nearly 80 per cent of respondents in the lowest social classes saying they belonged to the working class, compared with

TABLE 1.4.   Measured social class and subjective social class.

Registrar General's schema	% of respondents describing themselves as 'middle' and 'working' class			
	Middle	Working	Neither/ don't know	N
I	63	30	7	(150)
II	53	41	6	(816)
III (NM)	37	59	4	(944)
III (M)	18	78	4	(797)
IV	20	77	3	(708)
V	17	80	3	(201)
Goldthorpe's schema				
1	64	31	5	(363)
2	51	42	7	(547)
3	36	61	3	(880)
4	31	60	9	(258)
5	20	76	4	(202)
6	17	79	4	(413)
7	19	78	3	(960)

Note: Respondents are classified according to their own occupations.
Source: British General Election Study 1987.[1]

just over 60 per cent in the highest social class saying they belonged to the middle class.

Nevertheless, both the RG and Goldthorpe's classifications show a clear relationship between 'objective' class, as measured by these classifications, and subjective social class. Moreover, the two classifications give very similar results. Both show quite large differences between the top three (non-manual) classes identified, but no statistically significant differences between the bottom three (manual) classes. In other words, the relationship between objective and subjective class is not a linear one: the non-manual classes are fairly distinct from each other while the manual ones are more homogeneous.

We turn next to Table 1.5, which relates the two class schemas to educational attainment. As our measure of educational attainment we have taken an 'O'-level pass or its equivalent (a grade 1 at CSE or school certificate for respondents educated before World War 2).

As we can see, the RG scheme does discriminate rather better than Goldthorpe's, at least if we take the difference between the top and bottom classes as our criterion. (We should recall however our earlier caveat about the

TABLE 1.5.   Educational attainment by fathers' class.

Registrar General's schema	% of respondents obtaining an 'O'-level pass or equivalent	
		N
I	76	(141)
II	56	(684)
III (NM)	57	(327)
III (M)	34	(1515)
IV	30	(518)
V	19	(175)
Goldthorpe's schema		
1	71	(326)
2	67	(330)
3	55	(197)
4	37	(463)
5	43	(359)
6	31	(869)
7	26	(852)

Source: British General Election Study 1987.[1]

reliability of class V). Moreover, whichever scheme we use, we now see quite marked differences between the various manual classes: the manual classes may be homogeneous in their subjective class identity, but they are not homogeneous in educational terms.

Neither scheme, however, yields a linear relationship between class background and educational attainment. In the case of the RG scheme, it is class III(NM) that is out of line. We should note that Table 1.5 refers to the class that the respondent's *father* was in at the time the respondent was aged fourteen. This raises some interesting problems. First, such data on father's class will refer to the occupations that they held some years ago, when perhaps the nature of class III(NM) was rather different from what it is today. Secondly, men who were in class III(NM) occupations at the time they had 14-year-old children will be rather different from the typical incumbents of class III(NM), who would tend to be rather younger on average. The 'meaning' of a particular class should not, therefore, be regarded as fixed.

Let us now move on to consider Table 1.6 on voting behaviour.

This time it is Goldthorpe's scheme which tends to give the greater discrimination. And in particular Goldthorpe's schema brings out one very interesting finding, which is completely obscured by the RG scheme, namely the

TABLE 1.6.   Social class and voting behaviour.

| Registrar General's schema | % voting Conservative, Labour or, for other parties in 1987 | | | |
	Conservative	Labour	Other	N
I	52	12	37	(128)
II	52	19	29	(701)
III (NM)	54	21	24	(806)
III (M)	37	41	22	(679)
IV	32	45	23	(584)
V	25	52	23	(164)
Goldthorpe's schema				
1	61	12	27	(317)
2	47	20	33	(465)
3	52	24	24	(753)
4	64	17	19	(220)
5	39	34	27	(171)
6	31	46	23	(349)
7	30	48	22	(785)

Note: Respondents are classified according to their own occupations.
Source: British General Election Study 1987.[1]

propensity of the petty bourgeoisie to vote Conservative. Indeed, it proves to be the most Conservative-inclined of all the social classes. This brings out more clearly than anything else the non-hierarchical nature of Goldthorpe's class schema and the non-linear relationship between economic advantage and class action. The petty bourgeoisie is not particularly advantaged economically, and its support for the Conservative party almost certainly reflects its economic interest in free market policies. It is the market situation, rather than the actual occupations, of the petty bourgeoisie which surely explain their voting behaviour.

More generally, then, Tables 1.4, 1.5, and 1.6 show that there are somewhat different forms of relationship with class depending on which dependent variable is being considered. The tables also confirm the superiority of Goldthorpe's scheme in understanding class action and the more cultural nature of the RG scheme.

THE CHANGING SHAPE OF THE CLASS STRUCTURE

There are a number of respects in which the class structure may be changing. Firstly, certain occupations may be rising or falling within the class structure.

Considerable debate has been generated about the thesis of 'proletariani-zation' in particular. This holds that lower white-collar work has been progres-sively 'deskilled' by technological advances and should now be equated with manual work. While this may be true of specific white-collar jobs, it is less obvious that white-collar work *on average* has become proletarianized; new and more skilled white-collar jobs may also be created by new technology, and the net effect of the changes may instead be towards an upgrading of white-collar work (Braverman 1974; Stewart *et al.* 1980; Crompton and Jones 1984; Marshall *et al.* 1988).

Secondly, and much less controversially, there is the thesis that the shape of the class structure has been changing, with manual jobs declining in number and managerial and professional ones increasing.

Table 1.7 shows the changing distribution of the economically-active population over the post-war period. Occupations have been grouped into the Goldthorpe classes. (The census material on which Table 1.7 is based does not allow us to distinguish the higher and lower levels of the service class, and these have accordingly been grouped together.)

As we can see, professional and managerial workers have doubled in proportion from 12 per cent to nearly 25 per cent of the economically active population. The proportion of foremen and routine non-manual workers has also grown, while employers and self-employed have stayed more or less constant. The main losses have therefore been among the manual workers. In 1951, manual workers made up 62.3 per cent of the labour force; by 1981 this had declined to 45.4 per cent.

The economically active labour force is a different concept from that of the class structure. In studying the labour force, the unit of analysis is the individual, and the economically inactive (largely housewives and the retired) are ignored. Table 1.8 shows that there has been a major increase in women's participation in the labour force in the post-war period. This increase has been

TABLE 1.7.   Distribution of the economically active population, 1951 and 1981.

	1951	1981
Professional and managerial	12.0	24.8
Routine non-manual	16.3	19.3
Self-employed	6.7	6.4
Foremen/women	2.6	4.2
Skilled manual	23.8	16.0
Semi and unskilled manual	38.5	29.4
	99.9	100.1
	(22 514 000)	(25 406 000)

Source: Routh 1980, 1987.

TABLE 1.8. The female proportion of the labour force.

	% of each category composed of women in 1951 and 1981	
	1951	1981
Professional and managerial	29.1	34.0
Routine non-manual	56.3	76.7
Self-employed	18.2	19.7
Foremen/women	13.4	24.4
Skilled manual	15.8	12.5
Semi and unskilled manual	33.1	38.9
All	30.8	38.9

Source: Routh 1980, 1987.

particularly concentrated in routine non-manual work. As we noted above, our map of the class structure will depend upon which procedure we use for allocating individual respondents to classes: if married women are allocated to their husbands' class, then the size of the routine non-manual class is much reduced, for example.

However, the trends over time are much the same whichever procedure one uses. Table 1.9 uses the British Election Studies[2] to look at trends using the head of household procedure and the individual procedure respectively. Both procedures confirm the expansion of the service class and the contraction of the working class. Because of the concentration of women in routine non-manual work, the major differences concern this class. The individual procedure of allocating people to classes suggests that there has been a substantial increase in the size of class 3; the head of household procedure suggests that class 3 is much smaller and has not greatly increased in size.

ARE CLASS DIFFERENCES DECLINING?

Our final question is again a highly controversial one. Various writers have suggested that class differences might be declining, particularly in the field of political behaviour. For example, it has been suggested that increased affluence and the spread of home-ownership within the working class will have led to an 'embourgeoisement' of parts of the working class, particularly of more skilled workers. This has been seen as a major source of the recent Conservative election victories (Franklin 1985; Rose and McAllister 1986). On the other hand, the growth of unemployment may have led to increased poverty and deprivation and a greater polarization of society.

TABLE 1.9. The changing shape of the class structure.

Goldthorpe's schema	Respondents classified according to the head of household procedure	
	1964	1987
1	5.7	16.1
2	14.6	16.3
3	10.2	11.1
4	8.3	11.3
5	9.5	8.3
6	21.2	15.0
7	30.4	21.9
	99.9	100.0
N	(1535)	(3292)
	Respondents classified according to the individual procedure	
	1964	1987
1	5.0	13.0
2	13.8	16.4
3	14.5	19.1
4	7.3	9.2
5	8.5	6.5
6	19.0	12.3
7	31.9	23.6
	100.0	100.1
N	(1549)	(3300)

Source: British Election Studies 1964, 1987.[2]

We shall look at the same three dependent variables which were considered above, namely self-assigned class, educational attainment, and voting behaviour. We shall use Goldthorpe's class schema and, in order to maintain comparability with earlier data, will adopt the head of household procedure. We will once again use the British Election Studies as our database.

First, Table 1.10 looks at trends in subjective class identity.

As can be seen, there is a slight weakening of the relationship between 'objective' and 'subjective' class. Thus in 1964 68 per cent of class 1 defined themselves as middle class, falling to 65 per cent in 1987 while 15 per cent of class 7 defined themselves as middle class in 1964 rising to 17 per cent in 1987. Leaving aside questions of measurement error and comparability of the classes

TABLE 1.10.   Measured social class and subjective social class 1964–87.

Goldthorpe's schema	% of respondents describing themselves as 'middle' and 'working' class					
	1964			1987		
	Middle	Working	N	Middle	Working	N
1	68	20	(41)	65	30	(499)
2	51	40	(116)	49	47	(468)
3	49	49	(67)	35	58	(211)
4	46	43	(61)	32	61	(351)
5	18	76	(76)	21	76	(262)
6	12	85	(163)	16	80	(491)
7	15	81	(238)	17	81	(640)

Note: Respondents are classified according to the head of household's occupation.
Sources: British Election Studies 1964, 1987.[2]

over time, these are rather small changes and hardly suggest that there has been any major 'withering away' of class, at least at the subjective level.

Table 1.11 then looks at voting behaviour. In one respect, this table shows a very similar trend to Table 1.10: the trend for Conservative voting is very like the trend for middle class identity. Thus in 1964 64 per cent of class 1 voted Conservative, falling to 62 per cent in 1987 while 26 per cent of class 7 voted Conservative in 1964 rising to 31 per cent in 1987. The relationship between Conservative voting and 'objective' class has thus weakened, but only slightly.

There have however been much greater changes with respect to Labour and Liberal/Alliance voting. Labour's share of the working-class vote has plummeted by 20 points, most of these losses being transposed into Alliance gains. It is likely that a large part of these changes are due to factors that have nothing to do with class—Labour's disunity, the unpopularity of some of its policies among all social classes, and its ineffectiveness in government.

Finally, Table 1.12 looks at trends in class inequalities in education by comparing the educational qualifications of successive birth cohorts. Since the resulting cell sizes become rather small, we have collapsed the seven Goldthorpe classes into three; classes 1 and 2 are combined to form the service class, classes 3, 4, and 5 to form the intermediate classes, and classes 6 and 7 to form the working class.

As we can see, there has been a substantial increase over time, in all three classes, in the proportions obtaining an 'O'-level pass or its equivalent. The gap between the classes, however, has shown no secular tendency to decline. The simplest measure of the social class inequality is the difference in the percentages of respondents from service and working backgrounds obtaining

Anthony Heath

TABLE 1.11. Social class and voting behaviour 1964–87.

% voting Conservative, Labour, or for other parties in 1964 and 1987

Goldthorpe's schema	1964				1987			
	Conservative	Labour	Other	N	Conservative	Labour	Other	N
1	64	20	16	(81)	62	12	26	(440)
2	59	20	21	(204)	52	17	31	(408)
3	59	25	16	(142)	42	25	33	(180)
4	74	14	12	(112)	61	18	21	(289)
5	41	44	15	(124)	41	35	24	(222)
6	25	69	6	(283)	27	50	23	(427)
7	26	66	8	(394)	31	47	22	(529)

Note: Respondents are classified according to the head of household's occupation.
Sources: British Election Studies 1964, 1987.[2]

TABLE 1.12.   Class inequalities in education.

| Father's class | % obtaining an 'O'-level pass or its equivalent | | | | |
| | Birth cohort | | | | |
	1920–29	1930–39	1940–49	1950–59	1960–69
Service	44	58	68	87	83
Intermediate	21	31	48	59	78
Working	9	21	31	41	55
All	19	31	42	57	68
N	(509)	(518)	(621)	(643)	(618)

Source: British General Election Study 1987.[1]

'O'-levels. As we can see, this was 35 points in the 1920–29 birth cohort and it subsequently increases to 46 points in the 1950–59 cohort before falling back to 28 points in the 1960–69 cohort.

The percentage point difference is not however an entirely satisfactory measure for these purposes since it is subject to 'floor' and 'ceiling' effects, and a smaller gap between the classes is to be expected when overall levels of attainment are either close to the floor (as they were with the 1920–29 birth cohort) or close to the ceiling (as they were with the 1960–69 birth cohort). A measure such as the cross-product ratio (sometimes called the odds ratio) which is not subject to floor and ceiling effects is therefore more appropriate (see Fienberg 1980). Neither measure, however, suggests that there is any consistent trend towards smaller class inequalities.

Class differences of broadly similar magnitude seem to have persisted throughout the period covered by the available data. Tables 1.10 and 1.11 suggest that there has been a modest decline in the strength of the associations with subjective class and with voting behaviour, but this is not such as to call into question the value of the concept of class. On the other hand, Tables 1.7, 1.8, and 1.9 suggest that there have been major changes in the shape of the class structure over this period.

NOTES

1. The 1987 British General Election Study was a collaborative venture between SCPR and Oxford University, and was directed by Anthony Heath, Roger Jowell, John Curtice, Julia Field, and Sharon Witherspoon. It was funded by the Sainsbury Trusts, Pergamon Press, and the ESRC. The survey was a national probability sample representative of the registered electorate of Great Britain with an achieved sample size of 3826 respondents.
2. The 1964 British Election Study was conducted by David Butler and Donald Stokes and the data were made available to us through the ESRC Data Archive. We gratefully acknowledge their assistance and that of the original investigators.

REFERENCES

Acker, J.K. (1980). Women and stratification: a review of recent literature. *Contemporary Sociology*, **9**, 25–39.

Bland, R. (1979). Measuring social class. *Sociology*, **13**, 283–91.

Braverman, H. (1974). *Labour and monopoly capital*. Monthly Review Press, London.

Brewer, R.I. (1986). A note on the changing status of the Registrar General's classification of occupations. *British Journal of Sociology*, **37**, 131–40.

Britten, N. and Heath, A. (1983). Women, men and social class. In *Gender, class and work* (ed. E. Gamarnikow *et al.*), pp. 46–60 Heinemann, London.

Britton, M. and Birch, F. (1985). *1981 census post-enumeration survey*. HMSO, London.

Crompton, R. and Jones, G. (1984). *White-collar proletariat*. Macmillan, London.

Dodd, T. (1986). An assessment of the efficiency of the coding of occupation and industry by interviewers. *New Methodology Series 14*. OPCS, London.

Erikson, R. (1984). Social class of men, women and families. *Sociology*, **18**, 500–14.

Fienberg, S E. (1980). *The analysis of cross-classified categorical data*. MIT Press, Cambridge, MA.

Franklin, M.N. (1985). *The decline of class voting in Britain*. Clarendon Press, Oxford.

Goldthorpe, J.H. (1983). Women and class analysis: in defence of the conventional view. *Sociology*, **17**, 465–88.

Goldthorpe, J.H. and Hope, K. (1974). *The social grading of occupations*. Clarendon Press, Oxford.

Goldthorpe, J.H. with Llewellyn, C. and Payne, C. (1980). *Social mobility and class structure in modern Britain*. Clarendon Press, Oxford.

Hall, J. and Jones, D.C. (1950). The social grading of occupations. *British Journal of Sociology*, **1**, 31–55.

Heath, A., Jowell, R., and Curtice, J. (1985). *How Britain votes*. Pergamon, Oxford.

Jowell, R., Witherspoon, S., and Brook, L. (1988). *British social attitudes: the 5th report*. Gower, Aldershot.

Leete, R. and Fox, J. (1977). Registrar General's social classes: origins and uses. *Population trends*, **18**, 1–7.

Marsh, C. (1986). Social class and occupation. In *Key variables in social investigation*, (ed. R.G. Burgess), pp. 123–52. Routledge and Kegan Paul, London.

Marshall, G. (1988). Classes in Britain: Marxist and official. *European Sociological Review*, **4**, 141–54.

Marshall, G., Newby, H., Rose, D., and Vogler, C. (1988). *Social class in modern Britain*. Hutchinson, London.

Monk, F. (1978). *Social grading on the National Readership Survey*, fourth edition. Joint Industry Committee for National Readership Surveys, London.

Office of Population Censuses and Surveys (1970). *Classification of occupations 1970*. HMSO, London.

Office of Population Censuses and Surveys (1980). *Classification of occupations 1980*. HMSO, London.

Reid, I. (1981). *Social class differences in Britain*, second edition. Grant McIntyre, London.

Rose, R. and McAllister, I. (1986). *Voters begin to choose: from closed-class to open elections in Britain*. Sage, London.

Rose, D. Marshall, G., Newby, H., and Vogler, C. (1987). 'Goodbye to supervisors?' *Work, Employment and Society* 1, 7–24.

Routh, G. (1980). *Occupation and pay in Great Britain 1906–1979*. Macmillan, London.

Routh, G. (1987). *Occupations of the people of Great Britain 1801–1981*. Macmillan, London.

Stanworth, M. (1984). Women and class analysis: a reply to Goldthorpe. *Sociology*, **18**, 159–70.

Stevenson, T.H.C. (1928). The vital statistics of wealth and poverty. *Journal of the Royal Statistical Society*, **91**, 207–20.

Stewart, A., Prandy, K., and Blackburn, R.M. (1980). *Social stratification and occupations*. Macmillan, London.

Szreter, S.R.S. (1984). The genesis of the Registrar General's social classification of occupations. *British Journal of Sociology*, **35**, 522–46.

Wright, E.O. (1979). *Class structure and income determination*. Academic Press, New York.

Wright, E.O. (1985). *Classes*. Verso, London.

2

Social Class Mortality Differences

PETER GOLDBLATT

INTRODUCTION

Comparisons of mortality differences by social group in England and Wales are generally based on the Registrar General's social classes (P. Townsend *et al.* 1988). Their usage for this purpose dates back to the introduction by Stevenson of the prototype of the present schema in 1913 (Registrar General 1913). Initially he used the classification to analyse infant mortality information collected at the 1911 census but later extended its application to cover mortality at older ages by utilizing information collected from deaths in the period 1910–12 (Stevenson 1923). Since then, decennial comparison of deaths occurring to men of working ages around the time of each census with the corresponding population enumerated at census has provided the main index of social differences in mortality in this country.

The social class classification used for this purpose is occupationally based. The decision to select this particular axis appears to have related principally to pragmatic considerations such as convenience, precedent, and relevance to current policy issues (Leete and Fox 1977; Szreter 1984) rather than to any theoretical constructs (for example, a desire to group individuals according to either their relation to the means of production or the nature of their contract of employment). It built on a history of collecting occupational information at census and at death registration to compare mortality which dated back to the 1851 census (Registrar General 1855). In particular, when analysing occupational mortality in the period 1861–70, Farr used occupation to describe social position in an attempt to distinguish the effects of social and biological influences from the direct hazards of specific occupations (Registrar General 1864). To do this he used a social classification which placed most emphasis on grouping together broadly similar occupations in categories which corresponded to a loosely defined ordering of early industrial society (industrial workers, agricultural workers, domestic workers, etc.).

The link between these early analyses and the modern classification is described by Szreter (1984, 1986). A report produced by a British Association anthropometric committee (1883) proposed a more rigidly hierarchical classification of occupations, based on the conditions of life of the children of men following each occupation (Fig. 2.1). Szreter argues that this classification was predicated on an assumption of a strong relationship between achieved social status and inherited, natural differentials (Szreter 1986); that it provided a model which informed a 'nature versus nurture' debate in the early years of this century relating to differences in fertility and to 'highly undesirable cross-sectional differentials in physique and health' (Szreter 1984); and that it was this debate which provided the impetus for the introduction in 1913 of social class 'as a unidimensional graded hierarchy of occupations' in official statistics (Szreter 1984).

The leading proponents of its introduction did not, however, share a common position in this debate. For example, Mallet espoused broadly eugenic arguments (Austoker 1985), Newsholme environmental arguments (Szreter 1984), and Stevenson saw it as a means of empirically identifying and distinguishing causative factors (Registrar General 1913; Szreter 1984). Equally, the sole use of occupation as the basis of the schema adopted by Stevenson was not a foregone conclusion (Leete and Fox 1977; Szreter 1984). Prior to data collection in 1911, he favoured alternative indices of household position developed by Charles Booth, 'numbers of rooms, or of domestic servants, in relation to the number of persons in the family' (Stevenson 1910). However, by the next census he felt that these were even more flawed than occupation (Stevenson 1920). Similarly, uni-dimensionality was considered an unattainable goal in 1913, with three of the largest occupations (textile workers, miners, and agricultural workers) requiring separate identification as they lived in distinct communities whose fertility and infant mortality patterns did not accord with the social grading of the remainder of the population (Leete and Fox 1977; Szreter 1984). Their integration into the main classification was undertaken when data relating to 1921 were analysed (Registrar General 1927). Since then the classification used by the Registrar General has remained hierarchical, with each job assigned a class on the basis of an assumed level of skill or status (the emphasis placed on one or other of these changing over time).

The need to use cross-sectional methods for calculating mortality by class has meant that reliance has to be placed on occupational information which can readily be obtained on both occasions. It has also meant that these data are only available decennially, in supplements to the Registrar General's annual statistical volumes. As an index of mortality, class is thus sensitive both to discrepancies between contemporaneous sources, so called numerator–denominator biases (OPCS 1978), and to temporal changes in social and occupational distributions. Perception of the latter problem has lead to frequent modifications of the classification since its inception (Boston 1980). These are implemented by means of decennial assessments of the status (or skill) of current occupations, with the principal emphasis in this endeavour on grading

*Social Condition** Non-labouring classes		Labouring classes			
Nature† Very good	Good	Imperfect		Bad	Selected classes
Professional classes‡‡ (Upper and upper middle classes) 4.46 per cent.	Commercial class (Lower mid. classes) 10-30 per cent.	Labourers 47.46 per cent.	Artisans 26.82 per cent.	Industrial classes (Sedentary trades) 10.90 per cent.	Selected classes
Outdoor country§	Indoor towns	Outdoor country	Indoor towns	Indoor Towns	
CLASS I	CLASS II	CLASS III	CLASS IV	CLASS V	CLASS VI
Country-gentlemen	Teachers in elementary schools	Labourers and workers on agriculture	Workers in wood	Factory operatives	Policemen
Gentlemen-farmers	Clerks	" gardens	" metal	Tailors	Fire brigade
Officers of Army and Navy	Shopkeepers	" roads	" stone	Shoemakers	Soldiers
	Shopmen	" railways	" leather	etc.	Recruits
	Dealers in	" quarries	" paper etc.		
Auxiliary Forces	" drugs	Navvies	Engravers		Messengers§
Clergymen	" books	Porters	Photographers		
Lawyers	" wool	Guards	Printers		Industrial-Schools
Doctors	" silk	Woodmen	etc.		Criminals
Civil Engineers	" cotton	Brickmakers			
Architects	" foods				Idiots
Dentists	" drinks	Labourers, etc. on water			Lunatics
Civil Servants	" furniture	" Sailors			
Authors	" metals	" Fishermen			
Artists	" glass	" Watermen			
Teachers	" earthenware				
Musicians	" fuel, etc.				
Actors		Labourers, etc. in mines			
Bankers		" Coal			
Merchants (Wholesale)		" Minerals			

* Social condition; (influences of leisure, mental, and manual labour).
† Nurture; (influences of food, clothing, nursing, domestic surroundings etc.)
‡‡ Occupation; (influence of external physical conditions, exercise, etc.) Percentage of male population. including male children (Census of 1871).
§ Climatic and sanitary surrounding.

Fig. 2.1. Classification of the British population according to *Media,* or the conditions of life. *Source:* Szreter (1986) reproduced from the British Association Anthropometric Committee Final Report (British Association 1883).

working men (Szreter 1984). It is the combination of underlying social change and the associated classification changes which has produced considerable discontinuities in the size and structure of each class over time (Pamuk 1985).

The most recent *Decennial supplement on occupational mortality* (OPCS 1986) included data on mortality by class. However, analyses of these data were given less prominence in the commentary than was the case in the previous report (OPCS 1978). This reflected concern that the problem of numerator–denominator biases seemed to have worsened, possibly as a result of differential changes in class structure as recorded at census and at death. In this chapter, the evidence provided by recent studies of time trends is reviewed. These studies have been carried out in the context of a debate as to whether these trends reflect a variety of coincidental causal explanations, a single general explanation, or are purely an artefact of measurement. These differing theories are discussed below.

A weakness of using occupational class is argued by many to be the fact that it is a measure which can have direct relevance only to those in employment whose job is the major determinant of their social role. In summary it is a measure whose main focus can only be employed men at working ages; a limitation which is reflected in the emphasis of successive decennial supplements. It can only be applied by various indirect means to men out of employment (by reference to last main employment), to housewives (by reference to their husband), and to children (by reference to their fathers). These indirect solutions are frequently argued to create as many problems as they solve (Murgatroyd 1984) and to have serious theoretical limitations in terms of the lives of the individuals concerned (Osborn and Morris 1979).

A major problem faced by all critics of narrow, occupationally-based classification has been the difficulty of obtaining routinely collected data to evaluate alternative measures of social classification. Although individual data sources may collect a variety of alternative measures, few of these measures are collected in a comparable fashion across a range of data sources. To some extent the principal reason for this phenomenon is somewhat circular in nature; because occupational information is the only 'proved' method of standardized comparison, it is the only one collected on a wide variety of routine vital event records. In view of the costs and other difficulties encountered in the accurate collection and coding of occupationally-based data (Osborn and Morris 1979; Britton and Birch 1985), there is little enthusiasm for adopting more complex indices, which are unproved and have not withstood the test of continuous usage (Leete and Fox 1977).

In recent years new avenues for investigating these issues have opened through the use of longitudinal studies. With the repeated collection of questionnaire items at different points in time it is possible to evaluate comparability of each social measure over time, between generations and at successive stages in the lives of cohort members (Plewis 1985). When combined with prospective follow-up of mortality from the point at which class data are collected, it is in theory possible to both compare occupationally-based social class with alternative measures as predictors of event rates, free of numerator–denominator biases, and to evaluate whether the empirical basis of each measure seriously impedes its usefulness for time-dependent comparisons.

One study which combines many of these features (OPCS 1973) is the Office of Population Censuses and Surveys (OPCS) Longitudinal Study (LS). In this study records, starting with the 1971 census, are brought together continuously over time for a one per cent sample of individuals. Fuller details of the study are contained in the first report on mortality in the period 1971–1975 (Fox and Goldblatt 1982). Using this study, alternative indices of socio-economic differences in mortality can be formulated. The examination of differences for women and for those of both sexes who are beyond normal retirement age is also facilitated. Recent findings from this study confirm the existence of substantial mortality variation in each of these instances, affecting large groups at each end of the mortality spectrum. This chapter concludes with a discussion of the influence these results have on our understanding of class differences.

TIME TRENDS IN MORTALITY BY SOCIAL CLASS

Following the publication of the 1971 decennial estimates of mortality by social class (OPCS 1978) it was evident that reported differences were greater than previously recorded. This was highlighted in 1980 by the report of the

TABLE 2.1.   Mortality of men by occupational class (1930s–1970s) (standardized mortality ratios).

	Men aged 15–64					
	1930	1949	1959–63		1970–72	
Occupational class	–32	–53*	unadjusted	adjusted†	unadjusted	adjusted
I Professional	90	86	76	75	77	75
II Managerial	94	92	81	—	81	—
III Skilled manual and non-manual	97	101	100	—	104	—
IV Partly skilled	102	104	103	—	114	—
V Unskilled	111	118	143	127	137	121

* Corrected figures as published in *Registrar General's decennial supplement England and Wales, 1961: Occupational Mortality Tables.* HMSO, London, 1971, p. 22.
† Occupations in 1959–63 and 1970–72 have been reclassified according to the 1950 classification.
Source: The Black report, Table 7. (P. Townsend *et al.* 1988.)

DHSS working party on inequalities in health (P. Townsend *et al.* 1988), which pointed to the fact that differences in standardized mortality ratios (SMRs) had increased steadily since the 1931 *Supplement* (Table 2.1). However, the sensitivity of these data to numerator–denominator biases, to temporal changes in class distribution and to classification changes pointed to the need for caution in interpreting these trends. Since that report a number of analyses have been performed on these data to quantify the extent and source of this divergence so as to facilitate firmer conclusions. Preston *et al.* (1981) examined three summary indices of the data at each point in time from 1921–71 (the slope index of inequality, SII; the index of dissimilarity, ID; and the Gini coefficient, GC). Their analysis pointed to a consistent widening of differences since 1931 and they suggested that 'all groups have profited from mortality decline but higher status groups have profited more than most'.

Subsequent analyses have looked at two of these indices in more detail. Pamuk (1985) took account of the way in which reclassification moves occupational groups between classes by sequentially excluding the most severely affected occupations from the calculation of a class-based SII and by calculating an SII for 143 occupations which have remained relatively stable (Fig. 2.2). Both approaches confirm widening differences since 1931. However, the assumptions she makes in order to calculate and interpret these SIIs are considerable. In essence she assumes that each member of the population may be assigned a status ranking; that this is accurately measured by both occupation and class; and that mortality, after adjustment for age, is linearly related to this ranking. By this means she is able to convert occupation and class from categorical sub-divisions to interval scales and then regress mortality on the mid-points of these intervals. Moreover her assumption of an innate status ordering allows her to interpret classification changes as an accurate reflection of changes in relative position. Similarly she treats the class position of occupations in earlier years as equivalent to status ranking as assigned by the Goldthorpe scale in later years. The model is thus extremely rigid and it is, in theory, possible that changes in class structure and mortality patterns could have artefactually resulted in the trends reported.

In contrast to the SII, the index of dissimilarity requires few assumptions and simply indicates a notional, minimum number of deaths that would need to be reallocated between classes to remove inequality. Koskinen (1985) has used two versions of this to examine trends by cause of death. Both are based on the weighted sum of the absolute values of the difference between observed deaths in each class and those expected on the basis of SMRs for each of the other classes. The first expresses the notional reallocation as a percentage of deaths; the second converts this to a total number of deaths by taking account of age structure and numbers of deaths at each age. The index itself provides no indication of the direction in which deaths need to be reallocated. Where mortality for a specific cause increases or decreases steadily with class, this can

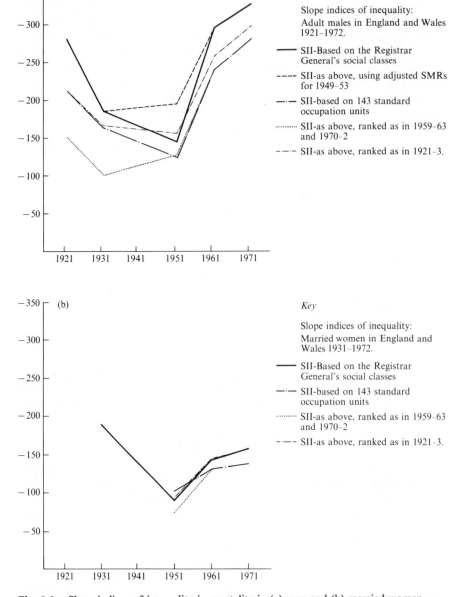

Fig. 2.2. Slope indices of inequality in mortality in (a) men and (b) married women. *Source*: Pamuk (1985).

be inferred without difficulty. However, if mortality does not vary monotonically with class the magnitude of the ID may be considerable but its interpretation is problematic. Thus variation in the ID over time cannot be interpreted without knowing the pattern of mortality at each time point. A second limitation of the ID is that even when it is expressed as a number of deaths, values for two separate causes cannot be added together. Thus disaggregation by cause must be treated with extreme caution.

Table 2.2 summarizes the changes in ID over time for causes of death which have, at some point, had a major numerical effect on mortality differences. This analysis identifies 1951 as the point at which all cause differences began widening after a period of narrowing. The table also provides a convenient summary of why this happened. The first half of this century saw the eclipse of respiratory TB and pneumonia as major killer diseases at working ages; thus although their social differentiation increased over time, the number of associated excess deaths in lower social classes decreased. Two diseases replaced them as the primary source of social differentiation, lung cancer and ischaemic heart disease. In the former case, Logan (1982) has shown how a major increase in rates for each class in the first half of the century was followed by the emergence of a clear social gradient from 1961 onwards as rates for higher classes levelled out and then fell while those for lower classes simply rose less rapidly (Fig. 2.3). This can largely be attributed to what is known about cohort changes in cigarette smoking and lung cancer (J. Townsend 1978a, b; Osmond et al. 1983).

The increasing pre-eminence of ischaemic heart disease in class differences may be traced to a combination of increasing rates over the period (Clayton et al. 1977) and of a reversal of class gradients (Marmot et al. 1978). While differences in smoking, diet, and leisure time activity are believed to have played a part in these trends (Rose and Marmot 1981; Cummins et al. 1981), changes in certification have probably contributed to this increasing role (Clayton et al. 1977; Marmot et al. 1978). It has also been argued that general improvements in living standards could have adversely affected heart disease rates among those who experienced impoverished childhoods (Forsdahl 1979; Barker 1987).

These analyses of decennial supplement data were all based on figures for censuses prior to 1981. Changes between 1971 and 1981 are more difficult to assess than usual as considerable changes took place in the size and structure of the labour force (Elias 1985; Goldblatt 1988). These were accompanied by a radical revision in the occupational classification used to derive class (Goldblatt 1988). The result appears to have been that the problem of numerator–denominator bias became worse and the size of classes at the extremes of the mortality spectrum even smaller. As a result direct class-for-class comparisons are misleading (OPCS 1986) and the appearance of a

TABLE 2.2. Index of dissimilarity as percentage, ID(%), and as number of deaths, ID(D). Causes of death for which ID(D) for men was greater than 500 in any year.

		Men					Married women			
		1922	1931	1951	1961	1971	1931	1951	1961	1971
All causes	ID(%)	3.5	2.6	3.4	4.6	5.3	3.7	2.5	4.3	5.0
	ID(D)	5499	3724	3552	4423	5012	2960	1184	1606	1929
Respiratory tuberculosis	ID(%)	7.4	9.6	9.3	10.7	19.4	10.0	10.5	9.4	14.2
	ID(D)	1919	1978	682	137	57	770	257	28	10
Cancer of lung	ID(%)	4.5	3.2	5.9	6.8	9.1	4.8	2.0	3.1	6.4
	ID(D)	42	48	521	765	1015	16	17	37	124
Ischaemic heart disease	ID(%)	—	-3.2	-2.9	2.6	3.3	5.3	3.5	5.7	8.0
	ID(D)	—	-395	-599	672	1028	342	145	273	450
Pneumonia	ID(%)	7.6	7.8	9.9	12.7	13.9	6.7	6.5	9.2	9.9
	ID(D)	1035	792	308	321	336	246	69	84	103
Bronchitis	ID(%)	—	—	11.3	13.1	13.4	—	11.8	12.0	13.5
	ID(D)	—	—	869	916	667	—	147	115	127

Source: Koskinen (1985).

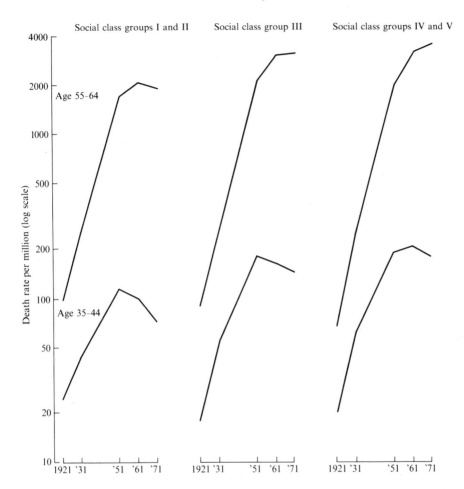

Fig. 2.3. Lung cancer death rates in men aged 55–64 and 35–44 by social class groups. *Source*: Logan (1982).

gross widening of differences illusory (*British Medical Journal* 1986). None the less two methods of examining trends since 1971 are available.

The first of these techniques simply involves aggregating classes into two broad groups, manual and non-manual, so as to minimize the effects of occupational mobility and classification change (Marmot and McDowall 1986). This suggests that the modest widening in class differences continues to be principally associated with lung cancer and heart disease (Fig. 2.4). Several factors previously associated with changes in the class distribution of these diseases continue to show wide class differences (OPCS 1984, 1985). However

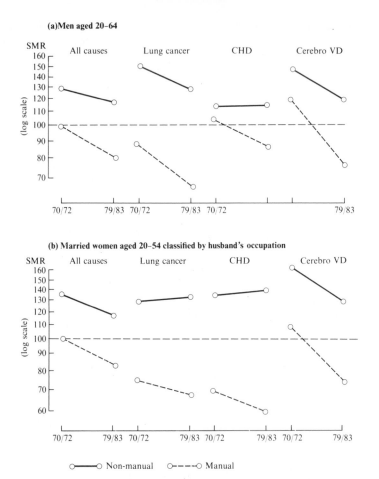

Fig. 2.4. Standardized mortality ratios for select causes of death in Great Britain 1970–72 and 1979–83 by social class. For each cause the SMR in 1979–83 is 100 for each sex. *Source*: Marmot and McDowall (1986).

this comparison of mortality levels may also have been affected by the rise in unemployment in the period (Marmot and McDowall 1986). The main reason for this is that numerator–denominator biases are particularly associated with movement out of the labour force, and this was markedly greater among manual workers over the decade (Goldblatt 1988; Fox and Shewry 1988). However, unemployment itself has also been related to an increase in mortality beyond that suggested by the class distribution of those affected (Moser *et al.* 1987).

The second method of looking at change is to move away from the use of

TABLE 2.3. Mortality of men by cause, age, time period, and social class in 1971: Longitudinal Study, 1976–83.

		Age at death					
		25–64		65–74		75 and over	
Cause	Social class group	1976–81 SMR	1981–83 SMR	1976–81 SMR	1981–83 SMR	1976–81 SMR	1981–83 SMR
All causes							
	I and II	75	78	79	81	82	81
	IV and V	114	115	108	111	109	109
	Non-manual	84	83	81	87	86	84
	Manual	103	107	103	105	107	107
Lung cancer							
	I and II	63	69	70	71	83	68
	IV and V	129	122	124	131	124	106
	Non-manual	69	75	70	74	77	83
	Manual	116	116	111	113	119	115
Ischaemic heart disease							
	I and II	77	72	91	84	85	91
	IV and V	114	110	96	114	101	99
	Non-manual	90	81	90	92	96	89
	Manual	104	109	100	104	100	102

Source: Based on Fox and Goldblatt (1986).

cross-sectional comparisons and to utilize prospective mortality data by class from the Longitudinal Study (LS) (Fox *et al*. 1985; Fox and Goldblatt 1982, 1986), which have recently become available. By retaining individuals in the same class as follow-up progresses, the importance of ensuring class comparability over time is considerably reduced. The method is however subject to several limitations. Health problems frequently affect the way in which people are categorized in any survey and this distorts mortality differences in a prospective study for a period shortly after data collection. Coding of class in the 1971 census magnified the effects of this on LS data in the early 1970s (Fox and Goldblatt 1982), so complicating the examination of trends (Fox *et al*. 1985). The study is also limited by covering fewer deaths than cross-sectional analyses. While longitudinal data ultimately provide a better tool for distinguishing age, cohort, and period trends (Plewis 1985; Osmond and Gardner 1982), this study as yet covers too short a span to realize this promise.

Table 2.3 shows how class differences among men in the study varied over the period 1976–83. While this confirms some widening of differences based on a manual/non-manual divide at working ages, it is less clear that this is also true at other ages or for other groupings of class. Equally re-basing recent figures to 1981 class (Fox and Goldblatt 1986) gives no credence to the idea that changes in class structure or classification can explain widening differences in cross-sectional data (other than through an effect on numerator–denominator biases).

Changes in heart disease differentials confirm a continuing and consistent widening of class differences. However, this no longer translates as readily to a widening gradient in all cause mortality because this disease has ceased to increase as a proportion of all deaths at relevant ages. Evidence for a continued increase in lung cancer differences is no clearer than that for all causes of death.

EXPLANATIONS OF A GENERAL PATTERN

This examination of trends in class differences highlights the effect that changes in social structure and disease patterns can have on the magnitude of mortality differences. A variety of explanations have been put forward to explain why these changes have persistently failed to alter the direction of class gradients.

Illsley (1986, 1987) has suggested that 'changes in the range of occupational class rates bear no necessary correspondence to changes in socio-economic equality'. Instead, divergence in death rates is consistent with 'changing class distributions associated with selective recruitment and loss'. Specifically, the steady erosion in the size of less advantaged classes (Table 2.4) has meant that classes with high death rates have formed an ever-diminishing segment of society. Since, in Illsley's view, class recruitment is always selective (in both health and socio-economic terms), the appearance of class differences is maintained as they are increasingly based on a small group whose occupation reflects their pre-existing ill health. The mechanism for generating inequalities

TABLE 2.4.   Distribution of economically active men by occupational class for England and Wales, 1931–1971.

Year	Occupational class					
	I	II	III	IV	V	All classes
1931	1.8	12.0	47.8	25.5	12.9	100
1951	2.7	12.8	51.5	23.3	9.7	100
1971	5.0	18.2	50.5	18.0	8.4	100
% change 1931–1971	+178	+52	+6	−29	−35	

Source: Adapted by Illsley (1986) from the Black report.

TABLE 2.5.   Example of the effect selective social mobility could have on mortality differentials.

(i) *Pre-mobility 'health' distribution: 1st generation*

'Health' (Predictor of death by age 40)

			'Health'			
		⌈100	Good	100 ⌉		
Rich	500	⎨300	Medium	300 ⎬	500	Poor
		⌊100	Bad	100 ⌋		

Probability of staying rich	'Health'	Probability of staying poor
1.0	Good	0.5
0.9	Medium	0.75
0.8	Bad	1.0

(ii) *Post-mobility 'health' distribution*

			'Health'			
		⌈150	Good	50 ⌉		
Rich	575	⎨345	Medium	255 ⎬	425	Poor
		⌊ 80	Bad	120 ⌋		

Mortality rates by age 40: Good = 0.02, Medium = 0.05, Bad = 0.08.

Mortality rate	Pre-mobility	Post-mobility
Rich	0.05	0.0463
Poor	0.05	0.0549
Differential (poor/rich)	1.00	1.187

Note: This example demonstrates that an increase in the rate of social mobility will increase social class mortality differentials provided that (a) health is a systematic social mobility selection factor and (b) the mobility process increases the average health level of the rich.
Source: Stern (1983).

through this type of selective social mobility is described by Stern (1983), using examples such as that shown in Table 2.5.

This interpretation of the data has been challenged by Wilkinson (1986*a*, 1987). He has argued that the indices used by Preston *et al.* (1981), Pamuk (1985), and Koskinen (1985) avoid contrasts between individual classes and focus on the social distribution of mortality across the whole population. He has also pointed out that little evidence exists to show that mobility is either becoming more common or that if upward mobility now predominates then this is more selective for health than downward mobility. While there is little historic data available to make the necessary comparisons, results from the LS provide an indication of what has happened in recent years.

Between 1971 and 1981 the number of middle-aged men who changed class was considerable (Fig. 2.5). Some of this was due to classification changes which masked some systematic individual change. However only a part of this

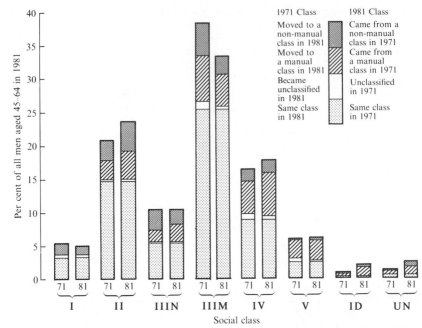

Fig. 2.5. Comparison of social class distributions in 1971 and 1981 for men aged 45–64 in 1981, showing the extent of class mobility. *Source*: Goldblatt (1988).

was associated with consistent upward mobility. Much of the reduction in the size of lower status groups at working ages can be traced to the ageing of earlier, larger cohorts and correspondingly lower levels of recruitment among younger men (Goldblatt 1988). The greatest difference in individual mobility by class is in the rate at which men become unclassified (Table 2.6). To the extent that mobility has a selective effect on mortality, it is the latter type of downward movement which can be demonstrated to be important (Fox *et al.* 1985). It tends to reduce apparent inequalities in the short term. Preliminary evidence (such as that provided by Table 2.3) suggests that changes between classes in the period 1971 to 1981 had only a small (and non-directional) effect on subsequent mortality. However, the effects of classification changes need to be quantified before firm conclusions can be drawn from these data.

In contrast to explanations which rest on artefacts of health selection, some researchers have suggested general explanations for the persistent pattern of mortality inequality (Marmot *et al.* 1984). These explanations attempt to link adverse social environment to illness, either through a cluster of recognized risk factors or through general susceptibility. As well as factors relating to

TABLE 2.6. Longitudinal Study* 1971 and 1981: change† in social class between 1971 and 1981 among men aged 45-64 in 1981 by economic activity in 1981 in England and Wales.

Economic activity in 1981		I	II	IIIN	IIIM	IV	V	Armed forces	Inad. described	Unoccupied	All
All men											
1971 distribution	%	5.4	20.6	10.5	38.3	16.4	6.1	0.6	0.9	1.3	100
Moved out	%	41	29	48	34	46	58	79	91	62	39
Moved in	%	33	43	48	21	54	59	9	224	105	39
Net change	%	−8	+14	0	−13	+8	+1	−70	+133	+53	—
1981 distribution	%	4.9	23.4	10.4	33.3	17.7	6.1	0.2	2.0	2.2	100
Employed men											
1971 distribution	%	6.1	22.2	11.0	38.4	15.8	5.0	0.6	0.6	0.4	100
Moved out	%	41	28	48	32	44	57	81	89	100	38
Moved in	%	32	45	48	22	55	66	5	71	6	38
Net change	%	−9	+17	0	−10	+11	+11	−76	−16	−94	—
1981 distribution	%	5.5	25.8	11.0	34.1	17.5	5.5	0.1	2.4	—	100
Men out of employment											
1971 distribution	%	2.4	13.8	8.3	37.6	18.8	10.9	0.4	2.2	5.6	100
Moved out	%	52	39	50	42	53	61	64	83	51	48
Moved in	%	40	32	46	22	51	45	36	394	133	48
Net change	%	−12	−7	−4	−20	−2	−16	−28	+311	+82	—
1981 distribution	%	2.1	12.8	8.1	29.8	18.5	9.1	0.3	9.1	10.2	100

Social class

*Members whose census records in 1971 and 1981 were matched and who were traced at NHSCR.
†All changes are expressed as a percentage of the 1971 size of the class in the corresponding economic activity category.
Source: Goldblatt (1988).

behaviour (smoking, diet, exercise), the former also include adversity trans-mitted from childhood such as height (Marmot *et al.* 1984; Power *et al.* 1986) and the effects of nutrition (Barker 1987). The source of childhood factors is usually ascribed to social deprivation (Forsdahl 1979; Barker 1987) rather than genetic factors (Marmot *et al.* 1984) but, in either event it argues against a pivotal role for health selection in creating health inequality (Power *et al.* 1986). General susceptibility has been proposed as a way in which stress, for example, may affect mortality over a range of causes and beyond its direct effect on health-related behaviour (Marmot *et al.* 1984).

ALTERNATIVE INDICES

Discussion of the existence of class differences in mortality and of changes in their magnitude often call into question the appropriateness of any occupational definition and of that used by the Registrar General in particular. While the theoretical limitations of the schema focus on the shortcomings of the classification in coherently ranking the prestige and status of occupations, it is the limitations of occupation as an index of way of life that dominate the debate. This pragmatic concern with social influences on the individual, rather than placing the individual in a set of formal relations can be traced back to Stevenson's criteria (1928) for judging the success of the classification, when looking back at the 1921 *Decennial supplement.*

Classification of individuals by income was not possible under present conditions in this country, though it had been employed on a very limited scale in America. Estimation of poverty by housing conditions was very unsatisfactory, as bad housing was only one of the handicaps of poverty, so that it was impossible to determine how far the excess of mortality associated with bad housing was due to poverty and how far to the direct effects of overcrowding, etc. Even if full details of income were available, these in themselves would not provide an ideal basis for classification, as it was probably the cultural associations of wealth which promoted longevity rather than wealth itself. . . . The method advocated for meeting the conditions to be considered was that of infering social position from occupation. By this means regard could be paid to (average) culture as well as income.

Since Stevenson opted for using occupational class, few alternative mortality analyses have been undertaken. The 1971 *Decennial supplement* (OPCS 1978) presented data for men by socio-economic group (SEG), shown in Fig. 2.6. This revealed considerable heterogeneity within classes III(NM) and III(M), with lower mortality among foremen, supervisors, and those working on their own account than among others in the same class.

Differences between SEGs in other classes were either less marked or varied by age. The main effect of this classification is thus to provide greater differentiation of some of the smaller groups at the centre of the social class

Non-manual classes

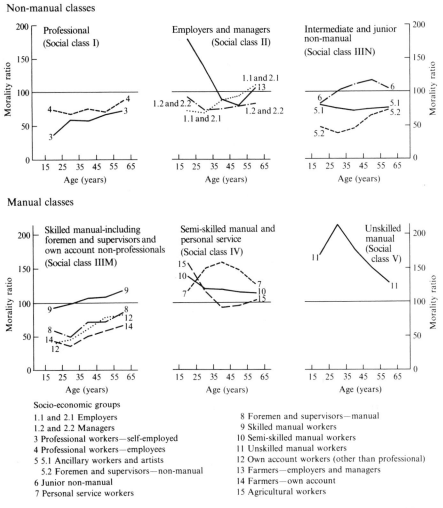

Fig. 2.6. Male mortality by socio-economic group and age, 1970–72. *Source*: OPCS (1978).

spectrum. In consequence, it appears there are groups within these classes which are together equal in size to social class 1 and which have comparably low mortality.

What the use of SEGs cannot do is to overcome the problems of an occupationally-based classification. In an attempt to derive an index based on individual variation rather than on that between socio-economic groups, Le Grand (1985) attempted to use the Gini coefficient to summarize variation in age at death at different points in time. This was in sharp contrast to Preston *et al.* (1981) who used it to measure variation in the discrepancy between observed and expected deaths for ranked social classes. Not surprisingly Le Grand's index decreased over time since, as Wilkinson has noted (1986*b*) it reflected nothing more than the change in the predominance of specific causes of death, each having a different age profile. As Illsley (1987) points out, it necessarily fails to measure socio-economic variation.

One question that must be asked is whether the changes in social structure, which have bedevilled social class comparisons, have so transformed society that Stevenson's rationale for selecting occupational class requires re-examination. Income data continue to be less readily available in this

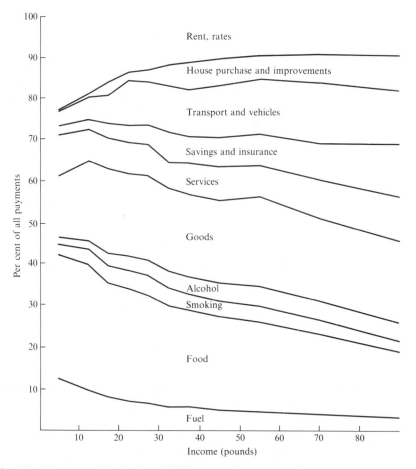

Fig. 2.7. Percentage distribution of different types of household expenditure and payments, by income, 1971. *Source*: Department of Employment (1972).

country than in America. For example, although sufficient data are now available to undertake limited correlation analyses (Wilkinson 1986c), the availability of census tract data in America facilitates small-area analyses of income level (Kitagawa and Hauser 1973). On the other hand, poor housing conditions (as measured by overcrowding or lack of basic amenities) have ceased to affect substantial numbers of people. Thus differences in housing can now be seen as relating more directly to wealth or poverty *per se*. Occupation, on the other hand continues to be as strongly associated with intrinsic hazards as with those resulting from way of life (Fox and Adelstein 1978).

If class is intended to indicate 'way of life' associated with different income levels then one approach is to focus on major items of income-related expenditure or wealth. Two household characteristics recorded by the census have been suggested as joint indices of both of these (Fox and Goldblatt 1982). After food, housing and transport were the most substantial items of household expenditure and outgoings which varied with income in 1971 (Department of Employment 1972), as indicated by Fig. 2.7. Within these categories, mortgage repayments and the running of a car were two of the most substantial elements which increased markedly as a proportion of outgoings. Correspondingly, Fig. 2.8 indicates how income distribution varied with housing tenure and household access to a car. Thus, these two items together provide not only a crude index of income but, equally importantly, of major ways in which income affects lifestyle and accumulation of personal wealth through expenditure patterns.

Table 2.7 contains data from the LS which compares mortality differences among men, based on these indices and on social class. Each index defines a consistent mortality gradient, but those based on cars and tenure are associated with more substantial groups at the extremes of the spectrum. Fig. 2.9 illustrates how these different indices interrelate. Lowest mortality is found among those with all the advantageous characteristics; failing to possess any one of them is associated with a higher level of mortality. At the other extreme of the socio-economic spectrum a slightly different picture emerges. Those with no car access and not in owner-occupied housing have comparably high levels of mortality, irrespective of class or of type of rented accommodation. Thus, while class assists in differentiating between advantaged groups, it is cars and tenure alone which identify mortality levels associated with disadvantage.

In the context of artefactual arguments put forward by Illsley (1986, 1987) and Stern (1983) (concerning the contrast between mortality rates in ever diminishing groups at the extremes of the spectrum), the relative size of the most advantaged and disadvantaged groups is of particular interest. Fig. 2.9 highlights the fact that cross-classification does not randomly distribute individuals because the three indices, though having separately identifiable effects on mortality, are strongly correlated with one another. Hence the most

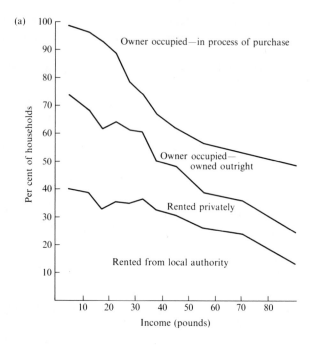

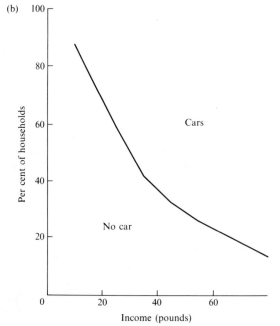

Fig. 2.8. Percentage distribution of households according to (a) housing tenure and (b) car access; by income, 1971. *Source*: Department of Employment (1972).

TABLE 2.7.   Mortality of men at ages 15–64 by alternative social classifications: Longitudinal Study, 1976–81.

Social classification	SMR	(Percentage of expected deaths)
Occupation-based		
Social class		
I	67	(5)
II	77	(20)
III (NM)	105	(10)
III (M)	96	(37)
IV	109	(17)
V	125	(7)
Other	189	(4)
Household-based		
Private households		
Tenure		
Owner occupied	85	(51)
Privately rented	108	(16)
Local authority	117	(31)
Car access		
Two or more	77	(15)
One	90	(50)
None	122	(33)
Non-private households	162	(2)
All men aged 15–64	100	(100)

advantaged group accounts for 22 per cent of expected deaths (SMR of 73) while those without access to a car and in some form of rented accommodation contribute 21 per cent (SMR of 123). This compares with 5 and 7 per cent for classes 1 and 5 respectively (for which the corresponding SMRs are 67 and 125).

WOMEN'S MORTALITY

Up to this point most emphasis has been given to class differences in male mortality. This reflects the limitations of the data available for women. This principally derives from an under-recording of women's occupation at death which has severely hindered successive decennial supplements. In 1908 Tatham identified a 'manifest need for definite information respecting occupational mortality among female workers' (Registrar General 1908), following an unsuccessful attempt to analyse data for 1900–2. In 1911 and 1921,

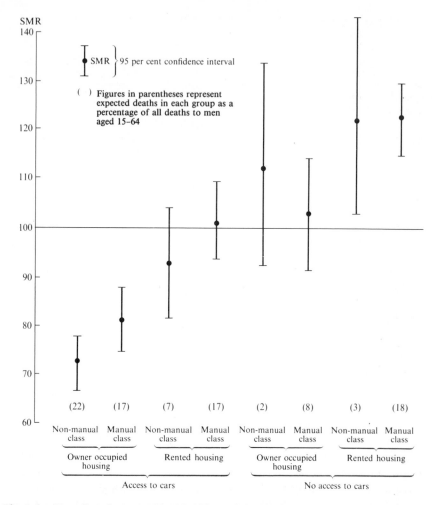

Fig. 2.9. Mortality of men at ages 15–64 by social class, housing tenure, and access to cars: Longitudinal Study 1976–81.

information on women's occupations remained largely unrecorded at death registration (Stevenson 1910, 1920), leading Stevenson to doubt 'whether this difficulty can ever be overcome . . . in view of the intermittent and transitory character of much female occupation'.

For married women, underrecording has continued to limit analysis (McDowall 1983; Roman *et al.* 1985), with registrars instructed to record husband's occupation unless the woman was in paid full-time employment either at death or for most of her life (OPCS 1978, 1986; Roman

TABLE 2.8. Mortality of women by occupational class (1961–71) in England and Wales (standardized mortality ratios).*

Occupational class	Women aged 15–64			
	Married		Single	
	1959–63	1970–72	1959–63	1970–72
I	(77)	(82)	(83)	(110)
II	83	87	88	79
III (NM)	} 103 {	92	} 90 {	92
III (M)		115		108
I V	105	119	108	114
V	141	135	121	138

*Based on *Registrar General's decennial supplement: 1961*, pp. 91, 503; OPCS, *Decennial supplement*, 1970–72, p. 211.
Source: The Black report, Table 7. (P. Townsend *et al*. 1988.)

et al. 1985). This practice is associated with the technique, introduced by Stevenson (1920), of analysing married women's mortality according to their husband's occupation. He saw this as a means of assessing whether mortality differences among men were due to the direct effects of occupation or to the social conditions implied by their occupation.

By contrast, for single women, the 1931 supplement (Registrar General 1938) recorded sufficient 'improvement in the comparability of statement in the death registers and on census schedules' to undertake analyses of mortality based on their own occupation (after making certain reservations concerning the unoccupied). Thus available *Decennial supplement* analyses comprise single women by their own class and married women by their husband's class (Table 2.8). Recent supplements have also included proportionate, cause-specific analyses of deaths for married women based on own occupation (OPCS 1978, 1986). The latter however provide no guide to overall levels of mortality. None the less, each of these analyses indicates that women's mortality is as socially differentiated as that of men.

Applying Stevenson's reasoning to the findings concerning married women's mortality by husband's class suggests that it is social conditions rather than occupation or health selection which account for class differences. However it has been argued that selective mating (Illsley 1986, 1987) and occupational endogamy (OPCS 1978) are important in interpreting differences in death rates. Both could contribute to this concordance of results. Similarly, marital selection (Berkson 1962; Kiernan 1988) and concern about numerator-biases limit the reliance that can be placed on corresponding analyses of single women.

Many of these caveats stem from the lack of any readily available method of

TABLE 2.9. Mortality of women at ages 15–59 by alternative social classifications: Longitudinal Study, 1976–81.

Social Classification	Marital status					
	Married		Widowed/ divorced		Single	
	SMR	(%)*	SMR	(%)*	SMR	(%)*
Occupation-based						
Own social class						
Non-manual	78	(23)	63	(2)	81	(6)
Manual	96	(22)	125	(2)	156	(3)
Other	106	(39)	139	(2)	226	(2)
Husband's social class						
Non-manual	73	(30)	—	(—)	—	(—)
Manual	108	(48)	—	(—)	—	(—)
Other	116	(6)	—	(—)	—	(—)
Household-based						
Private households						
Tenure						
Owner occupied	82	(43)	77	(2)	99	(5)
Rented	110	(37)	123	(3)	149	(5)
Car access						
One or more	84	(56)	84	(2)	85	(5)
None	122	(24)	115	(3)	161	(5)
Non-private households	108	(1)	143	(0)	214	(1)
All women aged 15–59	96	(84)	105	(6)	124	(10)

*Expected deaths in each category as a percentage of all deaths to women at ages 15–59.
Source: adapted from Moser *et al.* (1988*a*).

making direct comparisons between women which cover all marital status categories and are based on their own circumstances (Murgatroyd 1984; Osborn and Morris 1979). This problem can be overcome if information collected at census is examined prospectively. This avoids under-reporting at death and, by using alternative indicators discussed in the previous section, identifies factors as directly relevant to women's lives as to those of men (Arber 1987; Moser *et al.* 1988*a, b*). While housing tenure and access to cars differentiate women's mortality as well as they do for men, husband's class provides a lesser degree of discrimination (Table 2.9). The effect of own class varies by marital status; for single women marked differences are confirmed whereas for married women it principally serves to distinguish 'housewives' from those in paid employment at census (Moser *et al.* 1988*a*).

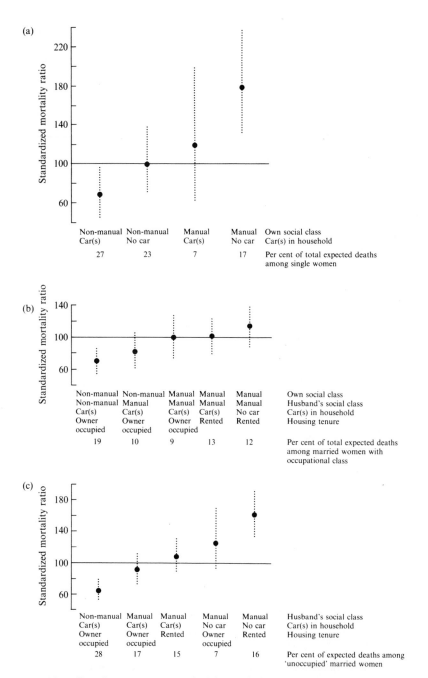

Fig. 2.10. Mortality of women at ages 15–59 by marital status, own and husband's social class, housing tenure, and access to cars, Longitudinal Study 1976–81: (a) single women; (b) married women with an occupational class; (c) 'unoccupied' married women. (Points are standardized mortality ratios and dotted lines are 95 per cent confidence intervals.) *Source*: Moser *et al.* (1988a).

TABLE 2.10. Mortality of married women at ages 15–59 by husband's class, parental status and own employment: Longitudinal Study, 1976–81.

Parental status	Husband's social class	Employment					
		Full-time		Part-time		Housewife	
		SMR	(%)*	SMR	(%)*	SMR	(%)*
Child aged under 17							
Non-manual		98	(3)	64	(6)	65	(11)
Manual		101	(6)	91	(10)	107	(17)
Children aged 17 and over							
Non-manual		63	(4)	72	(3)	81	(5)
Manual		96	(7)	83	(6)	136	(7)
Nulliparous							
Non-manual		84	(3)	46	(1)	85	(2)
Manual		105	(4)	109	(2)	144	(3)

*Expected deaths in each category as a percentage of all deaths to women at ages 15–59.
Source: adapted from Moser et al. (1988b).

Because occupational class fails to reflect many important aspects of women's, lives (Roberts and Barker 1986), it is of particular value to cross-classify women using all the characteristics identified in Table 2.9. As Fig. 2.10 indicates, this serves to identify large groups of women with very different levels of mortality. As in the case of men, the groups at the extremes of the mortality spectrum (Fig. 2.10) are also those whose socio-economic circumstances contrast most sharply (Moser et al. 1988a).

Among married women the 'intermittent and transitory' nature of employment, to which Stevenson referred, altered radically in the next 50 years (Hakim 1979). None the less, part-time employment and periods out of the labour force continue to characterize the employment pattern of many married women. However, as shown in Table 2.10, socio-economic differences persist (Moser et al. 1988b), irrespective of this variation in paid employment or of the main life-cycle factor influencing return to work (Newell and Joshi 1986).

DIFFERENCES AT OLDER AGES

Traditional emphasis on occupational class has had an even more inhibiting effect on the availability of data on mortality differences beyond normal retirement age than on comparisons of women's mortality at working ages. Concern in this instance focuses on the magnitude of numerator–denominator biases in cross-sectional data when recording of occupation is based on past, rather than current, occupation. Both the level and quality of recording are believed to deteriorate with increasing age and hence, in general, with time since last employment (OPCS 1978).

The approach taken in *Decennial supplement* analyses is to make proportionate comparisons using deaths shortly after retirement (up to age 74), but to make no use of census data or of deaths at older ages (OPCS 1978, 1986). By contrast, the emphasis on the prospective analysis of census characteristics in the LS obviates the need for caution concerning numerator–denominator biases (Fox *et al.* 1985) and transfers attention to the quality, validity, and interpretation of census data at older ages (Britton and Birch 1985; Fox and Goldblatt 1982).

As Table 2.3 shows, class differences in male mortality at older ages are as evident as those at working ages (Fox *et al.* 1985). While variation in cause-specific mortality in this table is related both to the contrasting life-experiences of different age cohorts and to factors affecting certification of cause of death at older ages, these are not issues which call into question the validity of gradients at older ages. The consistency of variation between individual classes is highlighted in Fig. 2.11.

While occupationally-based measures principally reflect the past experiences of retired people, household-based measures indicate current circumstances. Both are important, although the effect of health status on the living arrangements of old people raises questions about the validity of tenure or car access as socio-economic indices at these ages (Fox and Goldblatt 1982). However, as these reservations relate to ill-health related mobility among the

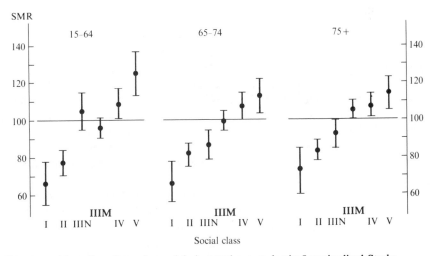

Fig. 2.11. Mortality of men by social class and age at death; Longitudinal Study 1976–81. *Source*: Fox *et al.* (1985).

TABLE 2.11.   Mortality of men by age and summary alternative social classifications: Longitudinal Study, 1976–81.

| | Age at death | | | | | |
| | 15–64 | | 65–74 | | 75 and over | |
Social Classification	SMR	(%)*	SMR	(%)*	SMR	(%)*
Occupation-based						
Social class						
Non-manual	84	(35)	81	(33)	86	(32)
Manual	103	(61)	103	(61)	107	(52)
Other	189	(4)	139	(6)	106	(16)
Household-based						
Private households						
Tenure						
Owner occupied	85	(51)	87	(53)	90	(55)
Rented	114	(47)	113	(45)	112	(42)
Car access						
One or more	87	(65)	86	(53)	87	(35)
None	122	(33)	115	(45)	106	(62)
Non-private households	162	(2)	151	(2)	129	(3)
All men aged 15 and over	100	(100)	100	(100)	100	(100)

*Expected deaths in each category as a percentage of all deaths to men in each age group.

elderly, they mainly affect differences shortly after the individual's characteristics are recorded at census (Fox *et al.* 1982). It is thus sufficient to allow a short period to elapse before making meaningful socio-economic comparisons.

Table 2.11 summarizes the main features of each measure for men in the period 1976–81. All identify differences in mortality at each age, although the proportion of men assigned a class decreases with age and lack of access to a car is as much a feature of age as of socio-economic circumstance. In this respect, housing tenure is, by contrast, relatively invariant with age.

Determining socio-economic differences among women at older ages is particularly problematic. Not only is occupation seldom recorded for these women (Table 2.12), but the majority are widowed and cannot be allocated a class on the basis of husband's occupation. Household-based measures are thus particularly relevant to the measurement of differences among elderly women. Car access is, however, even more strongly related to their age than it is for men. Hence most emphasis must be given to tenure in this instance. Although there is some narrowing of mortality differences by tenure for

TABLE 2.12. Mortality of women aged 60 and over by age group and alternative social classifications: Longitudinal Study, 1976–81.

	Age at death											
	60–74						75 and over					
	Marital status						Marital status					
	Married		Widowed/ Divorced		Single		Married		Widowed/ Divorced		Single	
Social classification	SMR	(%)*	SMR	(%)*	SMR	(%)*	SMR	(%)*	SMR	(%)*	SMR	(%)*
Occupation-based												
Own social class												
Non-manual	75	(10)	85	(4)	90	(5)	85	(2)	90	(3)	90	(4)
Manual	86	(13)	102	(6)	119	(3)	91	(2)	103	(4)	101	(2)
Other	101	(44)	125	(12)	140	(3)	93	(28)	106	(47)	104	(8)
Husband's social class												
Non-manual	81	(21)	—	(—)	—	(—)	86	(9)	—	(—)	—	(—)
Manual	98	(38)	—	(—)	—	(—)	93	(16)	—	(—)	—	(—)
Other	111	(7)	—	(—)	—	(—)	99	(6)	—	(—)	—	(—)
Household-based												
Private households												
Tenure												
Owner occupied	82	(35)	97	(10)	97	(6)	85	(17)	97	(24)	97	(7)
Rented	107	(31)	123	(11)	120	(4)	101	(14)	107	(26)	100	(6)
Car access												
One or more	80	(35)	92	(5)	88	(3)	82	(9)	99	(11)	80	(2)
None	109	(31)	117	(16)	114	(7)	97	(21)	104	(39)	101	(11)
Non-private households	185	(0)	210	(0)	175	(1)	128	(0)	163	(2)	117	(1)
All women	94	(67)	112	(22)	112	(11)	92	(31)	105	(54)	99	(14)

*Expected deaths in each category as a percentage of all deaths to women in each age group.

women aged 75 and over, considerable differences persist even in this oldest age group.

It would seem that the issue at older ages is not whether socio-economic differences exist, as there is strong evidence that they do, but that their measurement has proved exceedingly difficult. The indications are that these problems can now be largely overcome.

CONCLUSIONS

The main method of measuring social class differences in mortality, the Registrar General's occupational classification, is far from ideal. It is frequently criticized both for what it sets out to measure and for its failure to do even this effectively. None the less its conceptual simplicity has made it possible to collect data over a wide span of time and across data sources; comparison with other outcomes and influences on mortality are also facilitated.

In recent years this broad time-span has been utilized by researchers to suggest that these differences have widened since 1931. Although these conclusions are based on several alternative methods for summarizing trend data, none are free of sweeping mathematical assumptions. This has led some critics to suggest that health selection and the changing social structure have artefactually widened these indices of trend. They argue that these changes may even be associated with a real narrowing of differences. However their conclusion rests on hypothetical models, in place of mathematical assumptions about real data, and do not therefore warrant the same credence.

Two phenomena seem to argue for the persistence of mortality differences, by virtue of the fact that they are free of any mathematical assumptions and that their magnitude precludes all but the grossest of artefacts. The first is the change in the cause-specific contribution to differences, from predominance by infectious diseases to those most usually associated with lifelong health behaviour. Not only does this argue against a determining role for health selection, as selection would be required to produce comparable variation under two very different health scenarios, but it relates modern differences to behaviour which itself demonstrably varies by class. The implication of this is, however, that trends can only be understood in terms of the cohorts whose experiences are reflected in these changing patterns of mortality.

The second particularly revealing analysis relates to the examination of alternative socio-economic classifications for both men and women. This suggests that observed differences are not the product of using a particular social classification or of selecting only working men as the individuals whose occupation provides the basis for any class comparisons. Furthermore, by simultaneously using several axes to measure social circumstance we identify groups of men and women which not only have mortality levels as different as those recorded for social class 1 and 5 men but which comprise much larger proportions of the populations from which they are drawn. This indicates that the level of excess mortality associated with disadvantage is not something which relates only to an ever-diminishing group of individuals, as many authors have argued. It consequently suggests that this raised mortality is unlikely to represent artefacts either of changing occupational structure or (in view of the large numbers at the extremes) of selective health-related mobility.

NOTES

© Crown copyright is reserved.
1 The views expressed are not necessarily those of OPCS.
2 This work was supported by the MRC through grant number G8203453.

REFERENCES

Arber, S. (1987). Social class, non-employment, and chronic illness: continuing the inequalities in health debate. *British Medical Journal*, **294**, 1069–73.

Austoker, J. (1985). Eugenics and the Registrar General. *British Medical Journal*, **291**, 407.

Barker, D. J. P. (ed.) (1987). *Infant nutrition and cardiovascular disease*. MRC Environmental Epidemiology Unit: Scientific Report no. 8, Southampton.

Berkson, J. (1962). Mortality and marital status—reflections on the derivation of etiology from statistics. *American Journal of Public Health*, **52**, 1318.

Boston, G. (1980). Classification of occupations. *Population Trends*, **20**, 9–11.

British Association (1883). *Final Report of the Anthropometric Committee*. British Association Reports, London.

British Medical Journal (1986). Lies, damned lies and suppressed statistics (editorial). *British Medical Journal*, **293**, 349–50.

Britton, M. and Birch, F. (1985). *1981 Census post enumeration survey*. HMSO, London.

Clayton, D.Taylor, D., and Shaper, A.G. (1977). Trends in heart disease in England and Wales 1950–1973. *Health Trends*, **9**, 1–6.

Cummins, R.O., Shaper, A.G., Walker, M., and Wale, C.J. (1981). Smoking and drinking by middle-aged British men: effects of social class and town of residence. *British Medical Journal*, **283**, 1497–502.

Department of Employment (1972). *Family expenditure survey: report for 1971*. HMSO, London.

Elias, P. (1985). *Changes in occupational structure, 1971–1981*. Paper presented to Edinburgh Survey Methodology Group. Institute of Employment Research, University of Warwick.

Forsdahl, A. (1979). Are poor living conditions in childhood and adolescence an important risk factor for arteriosclerotic heart disease? *British Journal of Preventive Social Medicine*, **31**, 91–5.

Fox, A.J. and Adelstein, A.M. (1978). Occupational mortality: work or way of life? *Journal of Epidemiology and Community Health*, **23**, 73–8.

Fox, A.J. and Goldblatt, P.O. (1982). *Socio-demographic mortality differentials: longitudinal study 1971–75*. LS No. 1, HMSO, London.

Fox, A.J. and Goldblatt, P.O. (1986). *Have inequalities in health widened?* SSRU working paper 47, City University, London.

Fox, A.J. and Shewry, M.C. (1988). New longitudinal insights into relationships between unemployment and mortality. *Stress Medicine*, **4**, 11–19.

Fox, A.J., Goldblatt, P.O., and Adelstein, A.M. (1982). Selection and mortality differentials. *Journal of Epidemiology and Community Health*, **36**, 69–79.

Fox, A.J., Goldblatt, P.O., and Jones, D.R. (1985). Social class mortality differentials: artefact, selection or life circumstances? *Journal of Epidemiology and Community Health*, **39**, 1–8.

Goldblatt, P. (1988). Changes in social class between 1971 and 1981: could these affect mortality differences among men of working age? *Population Trends*, **51**, 9–17.

Hakim, C. (1979). *Occupational segregation*. Department of Employment Research Paper 9. Department of Employment, London.

Illsley, R. (1986). Occupational class, selection and the production of inequalities. *Quarterly Journal of Social Affairs*, **2**, 151–61.

Illsley, R. (1987). Occupational class, selection and inequalities in health: rejoinder to Richard Wilkinson's reply. *Quarterly Journal of Social Affairs*, **3**, 213–23.

Kiernan, K.E. (1988). Who remains celibate? *Journal of Biosocial Science*, **20**, 253–63.

Kitagawa, E.M. and Hauser, P.M. (1973). *Differential mortality in the United States*. Harvard University Press, Cambridge, MA.

Koskinen, S. (1985). *Time trends in cause-specific mortality by occupational class in England and Wales*. Proceedings of 20th International Population Conference, Florence. IUSSP, Liege.

Leete, R. and Fox, J. (1977). Registrar General's social classes: origins and uses. *Population Trends*, **8**, 1–7.

Le Grand, J. (1985). *Inequalities in health: the human capital approach*. Welfare State Programme, Discussion Paper 1. STICERD, London School of Economics, London.

Logan, W.P.D. (1982). *Cancer mortality by occupation and social class 1851–1971*. IARC SP no. 36/ OPCS SMPS no. 44, IARC/HMSO, Lyon/London.

Marmot, M.G. and McDowall, M.E. (1986). Mortality decline and widening social inequalities. *Lancet*, **ii**, 274–6.

Marmot, M.G., Adelstein, A.M., Robinson, N., and Rose, G. (1978). The changing social class distribution of heart disease. *British Medical Journal*, **ii**, 1109–12.

Marmot, M.G., Shipley, M.J., and Rose, G. (1984). Inequalities in death-specific explanations of a general pattern? *Lancet*, **i**, 1003–6.

McDowall, M.E. (1983). Measuring women's occupational mortality. *Population Trends*, **34**, 25–9.

Moser, K.A., Goldblatt, P.O., Fox, A.J., and Jones, D.R. (1987). Unemployment and mortality: comparison of the 1971 and 1981 Longitudinal Study census samples. *British Medical Journal*, **294**, 86–90.

Moser, K.A., Pugh, H.S., and Goldblatt, P.O. (1988a). Inequalities in women's health: looking at mortality differentials using an alternative approach. *British Medical Journal*, **296**, 1221–4.

Moser, K.A., Pugh, H.S., and Goldblatt, P.O. (1988b). *Inequalities in women's health in England and Wales: mortality among married women according to social circumstances, employment characteristics and life-cycle stage.* SSRU Working paper 57, City University, London.

Murgatroyd, L. (1984). Women, men and the social grading of occupations. *British Journal of Sociology*, **XXXV**, 473–97.

Newell, M-L. and Joshi, H. (1986). *The next job after the first baby: occupational transition among women born in 1946.* CPS Research Paper 86–3. London School of Hygiene and Tropical Medicine, London.

Office of Population Censuses and Surveys (1973). *Cohort studies: new developments.* SMPS no. 25. HMSO, London.

Office of Population Censuses and Surveys (1978). Occupational mortality. *Decennial supplement 1970–72.* DS no. 1. HMSO, London.

Office of Population Censuses and Surveys (1984). *General household survey 1982.* HMSO, London.

Office of Population Censuses and Surveys (1985). *General household survey 1983.* HMSO, London.

Office of Population Censuses and Surveys (1986). Occupational mortality. *Decennial supplement 1979–80, 1982–83.* DS no. 4. HMSO, London.

Osborn, A.F. and Morris, T.C. (1979). The rationale for a composite index of social class and its evaluation. *British Journal of Sociology*, **30**, 39–60.

Osmond, C. and Gardner, M.J. (1982). Age, period and cohort models applied to cancer mortality rates. *Statistics in Medicine*, **1**, 245–59.

Osmond, C., Gardiner, M.J., Acheson, E.D., and Adelstein, A.M. (1983). *Trends in cancer mortality 1851–1980: analyses by period of birth and death.* DH 1 no. 11. HMSO, London.

Pamuk, E.R. (1985). Social class inequality in mortality from 1921 to 1972 in England and Wales. *Population Studies*, **39**, 17–31.

Plewis, I. (1985). *Analysing Change.* Wiley, Chichester.

Power, C., Fogelman, K., and Fox, A.J. (1986). Health and social mobility during the early years of life. *Quarterly Journal of Social Affairs*, **2**, 397–413.

Preston, S.H., Haines, M.R., and Pamuk, E.R. (1981). *Effects of industrialisation and urbanisation on mortality in developed countries.* Solicited Papers, 2, Proceedings 19th International Population Conference, Manila. IUSSP, Liege.

Registrar General (1855). *Supplement to the fourteenth annual report of the Registrar General.* HMSO, London.

Registrar General (1864). *Supplement to the thirty-fifth annual report of the Registrar General.* HMSO, London.

Registrar General (1908). *Supplement to the sixty-fifth annual report of the Registrar General*, Part II. HMSO, London.

Registrar General (1913). *Seventy-fourth annual report of the Registrar General.* HMSO, London.

Registrar General (1927). *Decennial supplement, England and Wales, 1921. Part II: occupational mortality, fertility and infant mortality.* HMSO, London.

Registrar General (1938). *Decennial supplement: England and Wales 1931, Part IIa.* HMSO, London.

Roberts, H. and Barker, R. (1986). *The social classification of women.* SSRU working paper 46. City University, London.

Roman, E., Beral, V., and Inskip, H. (1985). Occupational mortality among women in England and Wales. *British Medical Journal,* **291,** 194–6.

Rose, G. and Marmot, M.G. (1981). Social class and coronary heart disease. *British Heart Journal,* **45,** 13–19.

Stern, J. (1983). Social mobility and the interpretation of social class mortality differentials. *Journal of Social Policy,* **12,** 27–49.

Stevenson, T.H.C. (1910). Suggested lines of advance in English vital statistics. *Journal of the Royal Statistical Society,* **LXXIII,** 685–713.

Stevenson, T.H.C. (1920). The fertility of various social classes in England and Wales 1850–1911. *Journal of the Royal Statistical Society,* **LXXXIII,** 401–44.

Stevenson, T.H.C. (1923). The social distribution of mortality from different causes in England and Wales. *Biometrika,* **XV,** 382–400.

Stevenson, T.H.C. (1928). The vital statistics of wealth and poverty. (report of a paper to the Royal Statistical Society). *British Medical Journal,* **i,** 354.

Szreter, S.R.S. (1984). The genesis of the Registrar General's social classification of occupations. *British Journal of Sociology,* **35,** 522–46.

Szreter, S.R.S. (1986). The first scientific social structure of modern Britain 1875–1883. In *The world we have gained,* (ed. L. Bonfield, R.M. Smith, and K. Wrightson). Blackwell, Oxford.

Townsend, J. (1978a). Smoking and class. *New Society,* **43,** 709–10.

Townsend, J. (1978b). Smoking and lung cancer: a cohort data study of men and women in England and Wales 1935–70. *Journal of the Royal Statistical Society* (A), **141,** 95–107.

Townsend, P., Davidson, N., and Whitehead, M. (1988). *Inequalities in health: the Black report and the health divide.* Penguin, Harmondsworth.

Wilkinson, R.G. (1986a). Occupational class, selection and inequalities in health: a reply to Raymond Illsley. *Quarterly Journal of Social Affairs,* **2,** 415–22.

Wilkinson, R.G. (1986b). Socio-economic differences in mortality: interpreting the data on their size and trends. In *Class and Health,* (ed. R.G. Wilkinson). Tavistock Publications, London.

Wilkinson, R.G. (1986c). Income and mortality. In *Class and health,* (ed. R.G. Wilkinson). Tavistock Publications, London.

Wilkinson, R.G. (1987). A rejoinder to Illsley. *Quarterly Journal of Social Affairs,* **3,** 225–8.

3

The Demography Of Social Class

D.A. COLEMAN

INTRODUCTION

Demography is the formalization of the risks or chances involved in belonging to a particular society: risks of dying and ill-health, risks of marrying and giving birth, risks of migrating, or of living in a large or small household—many of the characteristics of daily life as well as of its beginning and end.

The way people live and die can differ radically between different countries, different races and ethnic groups, and over periods of time. Demography provides a way of measuring these differences. But within all but the least complex of societies, there is also stratification by rank, income, or prestige (see Bendix and Lipset 1967), even in egalitarian ones (Rosenfeld 1951). In earlier times and in more traditional societies these are kept rigid by inheritance and in India fixed into castes through the joint reinforcement of religion, ethnicity, and status. The position is made much more complex, particularly in industrial societies, by substantial mobility between classes; both between generations and within them, at least in part on meritocratic grounds. But birth still complicates the picture. The children of the rich or of the professional classes in industrial societies have a particularly low risk of falling down the social ladder. So do the children of the *nomenklatura* in the Soviet Union, despite its being, as a Gorbachev aide claimed at a White House dinner in 1988, 'a proletarian society in which there is no place for bow ties'.

Social class, or some rank order based on occupations, the skill required to perform them, and the prestige accorded them, affects demographic risks in all Western societies of which we have knowledge; even those where social class is believed not to exist or to have been abolished (Andorka 1978; Fussell 1984; Jones and Grupp 1987). How it does so is still a matter for debate. This chapter will describe some social class differences in fertility and marriage. It will chart the origins of the present pattern of class differences, discuss their import, and speculate on their future. Some examples will be given from a variety of societies but most will be taken from the UK. No other country has so comprehensive a routine recording of data related to social groupings, and nowhere else has so much effort been made to devise systems, official and private, to capture social differences statistically (Reid 1981; Wrong 1958). Mortality is considered in Chapter 2. Throughout, the reader must keep in mind the difficulties of such classifications analysed by Anthony Heath in Chapter 1. His caveats will not be repeated here, although the section that follows draws attention to a gap in the 'official' class system which may be of demographic significance.

POPULAR AND OFFICIAL NOTIONS OF CLASS

Throughout this Chapter, most observations will be based upon the Registrar General's social class schema and related scales such as socio-economic group (OPCS 1980). The former was originally devised for use with the census of 1911 following pioneering work by William Farr (Stevenson 1920, 1923, 1928; Leete and Fox 1977; Boston 1984; Szreter 1984). Its lack of thorough statistical validation has long been a source of criticism. But the scales which have been proposed to replace it, when condensed to a usable number of categories, all bear a strong resemblance to its categories. Some earlier replacements, like the Hall–Jones schema (Hall and Jones 1950), have disappeared without trace. Use of more modern scales (e.g. Goldthorpe and Hope 1974) by workers other than their own creators has been rather limited.

The official social class currency originally minted by the Registrar General and issued by the OPCS (Office of Population Censuses and Surveys) still dominates the market. Official statistics—and their official categorizations—are still the chief source for academic enquiry and business and the related 'ABC1C2DE' scale of purchasing power is widely used in marketing. Despite its problems, the Registrar General's scale clearly does account for a high proportion of the observed variation of almost all demographically interesting variables. It has been suggested that the ideal might be to start with the behaviour being studied and build up categories from those sub-groups most similar in their behaviour. But that might multiply classification schemes to excess, impede comparison and risk circular argument.

A further problem is that none of these official or academic social class scales bears much relationship to some important popular conceptions of social class. This is because they do not admit the existence of, for example, an 'upper' class, an 'upper-middle' class or a 'ruling' class. Yet the aristocracy still lives, and its past records have provided important data for historical demography (Hollingsworth 1964). At least until recently, the notion of a 'ruling' class reflected reality (Guttsman 1969). An 'upper', or at least an 'upper-middle' class, and other elite groups clearly exist (Bottomore 1966; Scott 1982; Marwick 1986), despite its lack of official labelling (although the term was discussed in official writings when the social class scales were being elaborated earlier this century—see Boston 1984). It is that small minority of people, readily identifiable by dress, speech, and taste, who dominated literary attention up to the Second World War (in the works of Waugh, Stewart, Olivia Manning, Agatha Christie, and the rest, now known to a wider modern audience through television) but who have remained invisible to sociologists. On the whole, social scientists in Britain have had more empathy with the council flat than the country house and have felt that their role has been more to change the world than just to study it, and that it has been appropriate to study the demography of the aristocracy and gentry only when they are good and dead.

The public, the media, and their candidate members retain a lively interest in categorizing privileged minorities (Furbank 1985), witnessed by their successful promotion and marketing as 'Sloane Rangers', 'Yuppies', and everyday stories of media folk in cartoon strips. Even though the term is begotten by narcissism out of media hype, many are pleased to be identified as Sloane Rangers (Barr and York 1982) through conspicuously understated clothing and other visible affectations. They are the lineal descendants of an earlier bout of self-conscious labelling in the 1950s. Then untrained but acute participant observers like Nancy Mitford (1956) gave field guides for identification based mostly on language, which in Britain has occupied an exceptionally important position in popular, as opposed to academic systems of class labelling at least from the nineteenth century. This certainly connects with social reality through the unusual British school system and public school norms of speech. As Shaw remarked in 1912, 'It is impossible for an Englishman to open his mouth without making some other Englishman hate or despise him' (see Honey 1989).

By contrast the deracinated and ambiguous Yuppy has not travelled well from his American homeland. It is not clear what the acronym stands for, or what categories of jobs or tastes it implies except that they are overpaid, dubious, vulgar, and probably transient. Some US labels for modern couples even allude to their supposedly characteristic demographic behaviour ('Twinkies': two incomes, no kids), and some popular accounts claim distinctive Yuppy behaviour too; for example that exhausted Yuppy cohabitees have not time nor energy for sex. All this greatly displeases the 'Masses' (Middle Aged, Socially Static, Ever so slightly Scared).

The privileged minorities which live behind this smoke-screen of gossip are defined more by ancestry or by acquired taste and aspiration than by occupation. Such persons can inhabit social class 1 (professionals) or 2 (as farmers of varying degrees of grandness or even as businessmen), or have no occupation at all ('independent means'), but probably not elsewhere in the social class or income scale before losing all pretension to gentility.

Reminding readers of their existence is not a frivolous exercise. Their behaviour still attracts interest out of all proportion to their number. Their behaviour or supposed behaviour has been, and remains, the chief role model in British society. Their standards have indeed been taken to represent those of all British society (Klein 1965a). No data have been compiled on the demography of the upper social strata; their family size or age at marriage. Such data are only available for occupational elites such as the professions or for such dubiously and marginally high-status groups as senior academics (Hudson and Jacot 1971). The British élite may have distinctive demographic characteristics. The minority which this social group comprises of social class 1 may be responsible for giving it most of its distinctive demographic features. Their example may therefore be important, especially when their behaviour may be more influenced by their own self-image and ideology than that of others in the same job but of different social origins.

ORIGIN OF CLASS DIFFERENCES IN FERTILITY AND MARRIAGE

Social status differences in fertility, marriage, mortality, and migration are some of the best established relationships in social science. The 'higher fertility of the lower classes' has become, in a widely quoted source, almost a sociological truism (Wrong 1967). But their present form, though familiar, has been changing fast and may well be transient. Despite the recent expansion of historical demographic studies on fertility and family life, social differences in fertility and marriage seem to have taken a back seat. Occupation is only occasionally noted in the records of the parish register period (1538–1837) and then usually for purposes of identification rather than intellectual curiosity. Data from the censuses of the earlier nineteenth century have not been studied. So it is impossible to make confident generalizations about past British society as a whole.

The aristocracy are one of the few classes of society about whom much can be said (Hollingsworth 1964); all aspects of their lives are well documented. From the Middle Ages until the beginning of the early modern period the aristocracy had more children than average, primarily because they married earlier (Stone 1965, 1977). But from the seventeenth century onwards they adopted the late marriage already typical of the rest of the population, with correspondingly reduced fertility. In the eighteenth century they were noted for the lateness of their marriages and for evading it altogether. What little is known about the rest of society suggests that social class differentials in Britain and elsewhere (Knodel 1970) seem to have been quite modest before the decline of fertility which began in Britain with the marriage cohorts of the 1870s.

In many traditional societies outside Western Europe, where family planning is not practised, there may be few status differences in fertility, as for example in traditional rural China earlier this century (Notestein 1963). Elsewhere, higher-status women in rural areas often have higher fertility than those of lower status, for example in Nag's (1968) study of Bengali villages. In rural Bangladesh higher status may bring higher fertility. The better health of higher-status women protects their fecundity, they can breast-feed for a shorter time, which shortens birth intervals, their higher income permits them to avoid economic activity and instead fulfil their society's ideal role of women, with purdah enforced, inegalitarian role relations, and high coital frequency (Stoeckel and Chowdhury 1980). Similar positive relationships of status and fertility can be found elsewhere, examples being in rural Iran in 1974 (Ajami 1976) and in Java where there was a difference of between one and two children between upper and lower rural income groups, thanks mostly to differences in marital disruption, postpartum abstinence, and fecundity (Hull and Hull 1977). But the evidence is far from unanimous; contrary examples can easily be found (Bulatao and Lee 1983; Rogers 1984).

Some of these differentials may be transient too. High fertility in high-status groups may be eroded by the different diffusion of modern ideas from the top down. The early stages of modernization often encourage a transient increase in overall fertility and may do so particularly in low-status groups which previously suffered poor health. The existence of strong social stratification itself is regarded as a potential impediment to the spread of low fertility; obstacles to social mobility may prevent the advantages of low fertility being realized (Safilios-Rothschild 1982).

In traditional rural Europe associations of fertility with farm size and with ownership versus tenancy can be found. In pre-war Poland, for example, peasants with more land had bigger families than those with less land or landless labourers (Stys 1957), primarily because they could marry earlier. A study of a Normandy parish from 1901–1975 showed that owners of farms had later marriage and lower fertility (2.9 children) than tenants of farms (3.7 children) or non-farm workers (4.3 children). Here, fertility differences depended on position in the labour market and the importance of capital accumulation to the family fortune (White 1985). The two studies are not necessarily contradictory.

But if the élites tend to live in towns, the process of modernization is more advanced, or if women have little or no economic role in any class, then the differentials can resemble the expected Western pattern. Female participation in the non-agricultural labour force is particularly likely to depress fertility (Goldstein 1972; Wat and Hodge 1972). For example, a study in rural Iran showed inverse relationships between fertility and education, economic status and 'modernity', with higher-status women wanting fewer children, marrying later, and using contraception more. There are two important dimensions within the society; high/low status, and traditional/modern, with different effects on fertility (Delvecchio Good et al. 1980).

Some modern theorists (e.g. Becker 1981) insist that the underlying pattern in modern societies is that higher-income families have more children, as indeed Malthus expected would follow from the principle of prudential restraint. According to this view, the preponderance of large families among the poor and ill-educated in industrial society only persists today because their demographic transition is not yet complete. Their higher fertility is owed to unwanted births from poor family planning, part of a generally inadequate control over their circumstances. At higher status or income levels the educational effect upon contraceptive practice becomes saturated; that is, even higher status or income cannot bring substantial or commensurate improvements in contraceptive knowledge or practice. Further up the income scale the effect of income on fertility becomes positive, with additional children being considered once material needs are satisfied. This view, which is not without its critics (Blake 1968; David 1986) assumes that children have no important economic utility to their parents, and that they are in some respects competitors for family income with the satisfaction of material needs.

By the early post-war years there was some evidence for this view in the USA, where the relation of fertility to status was positive among families which planned the number and spacing of their children and U-shaped where only number was planned, with middle-status groups having the lowest fertility. Only among non-planners was low status clearly associated with high fertility (Kiser and Whelpton 1950). Later studies have not contradicted this finding, although differentials in general have diminished (Westoff *et al.* 1963). Most simple analyses of fertility with income, with only a few categories, still show an overall negative relationship (Wineberg and McCarthy 1986).

Such positive relationships may also be detectable in Sweden. There, in the 1970s, a U-shaped distribution gave the same average fertility among the highest and lowest income families. Above average income, fertility increased with rising income; below it, the reverse was true. There was a positive relationship between income and lower-order births with one-child and childless families being particularly rare; beyond the fourth child, the relationship was negative (Bernhardt 1972).

Any theory which attempts to show why some groups in society will want, and produce, more children will have to resolve the problem of why any parents in modern society should want any more children at all—one of the most difficult questions in modern demography. We know about their cost in money and time to both husband and wife, we know less about how decisions are made in modern families concerning children, how far this power may have devolved to the wife, and the proportion of children born despite or outside the decision-making process. The costs of children in modern society, including those arising from the lost opportunities for the wife to earn, are considered below (see p. 83).

Almost all discussion on low fertility in developed societies ignores social differences (Davis *et al.* 1986). And indeed it is true that national differences in average fertility seem to owe little to the variation of social status or income within societies; thus despite substantial differences in their pattern of social and economic inequality, Britain, France, and Japan have almost identical Total Fertility Rate (TFR) (1.8) and expected family size (2.2); so do Spain, Canada, Bulgaria, and the USSR. But it must be important that within these societies different social groups have quite different fertility. The balance between these sectors is not permanent, and the outcome of change may effect aggregate fertility.

THE HISTORY OF SOCIAL GRADIENTS IN FERTILITY

The adoption of family planning by some groups in society before others, in almost all Western societies in the nineteenth and early twentieth centuries, has created a gradient of higher family size by lower social class which has since come to be regarded as normal. In this country the conventional picture is derived mostly from the 1911 census and the 1946 fertility census (Glass and Grebenik 1954). Different social classes are regarded as having gone through the demographic transition at different times (Innes 1938; Lewis-Faning 1949; Glass and Grebenik 1954) more or less in rank order. The conventional view is that the motivation was the threat posed to their prosperity in the new economic circumstances by their traditional family size; the facilitators were the knowledge and literacy, and the impediments the social isolation, domestic economy, and adherence to tradition of the groups concerned (Banks 1954).

The simplicity of the idea that the middle class had begun to limit family size while natural fertility still prevailed among most manual workers has been challenged (Woods and Smith 1983) through evidence of substantial regional differences; but other research supports the original generalization, although with a rearranged rank-order.

Family limitation can first be inferred (from the decline in their marital fertility) among the families of the professions, particularly the clergy, medical men, and lawyers. After them, families in social class 2, civil servants, school teachers, accountants, and the like. In the middle classes, business men and farmers began to limit their families last.

In general, working-class families reduced their families later, but among them there were some important differences. In general, skilled artisans adopted small family size before labourers; occupations where married women customarily worked (e.g. textiles) before others, while agricultural labourers (Hinde and Garrett 1988) and coal miners, despite the skills and high income of the latter, were the last to follow, preserving high fertility into the twentieth century. These were the occupations most cut off from the rest of society, and in mining communities women had the least opportunity of joining the local workforce in any capacity and preserved a traditional, segregated marital role (Dennis *et al.* 1956).

Servants were essential to run a large house and family, so that the middle-class mother should be free for her prime role of managing her house, not labouring in it, and caring for her children's development. Even better paid working-class households would employ some living-out help. And their wages increased by 30 per cent over the period through competition with industrial wages. At that time almost no middle-class married women worked, although some working-class mothers did—notably in textiles. Domestic and personal service accounted for between 12 and 16 per cent of the entire workforce from 1801–1911, the peak in numbers being 2.6 million in 1911, although by then their percentage in the workforce had fallen from its 1891 peak of 16 per cent (Deane and Cole 1969). Then, domestic servants comprised the biggest single employment group outside agriculture. After the First World War and especially after the Second, numbers in domestic service fell precipitately to 500 000, or 2.2 per cent of the occupied population, in 1951. This very large workforce in service created fertility differentials at the lower end of the social scale too, because it was very difficult for persons employed in service to marry (Hinde and Garrett 1988).

According to the 'classical' view (Banks 1954) the old fertility regime was made obsolete by the increased costs of children, especially the emphasis on education in a modernizing society; and change was precipitated by the failure of economic growth in the 'Great Depression' after 1873. A general transfer of profits and salaries to wages tempted domestic servants into other employment, who took with them one of the props of the old fertility regime, and accordingly pushed up the wages of those who remained.

Positions in the services, professions, and the civil service (now much expanded) now required some evidence of ability and knowledge. Girls' education was taken more seriously too; all girls' public schools date from after 1870. Preparatory schools proliferated to replace tuition at home and to take middle-class children away from the Board Schools after the influx of working-class children which followed the introduction of compulsory education in 1876. All these were new costs (see Hair 1972).

The lower-middle class also saw the need for their children to be better educated, if only to keep them out of manual work, even though the multitude of clerks' jobs, then usually held by men, were often less well paid—£70 per year—than much manual work. Generally speaking, shopkeepers, clerks, junior teachers, and businessmen limited their families later than the professional classes—but possibly because of the greater hold of traditional and religious influences on their behaviour rather than because their need was any the less (Woods and Smith 1983).

Subsistence, not style, was the preoccupation in working-class household economy. Whether children's income from industrial employment ever really made a serious contribution to the working-class family budget is still controversial (Macfarlane 1986). But the employment open to children and the age and the hours at which they could work were progressively curtailed by legislation throughout the nineteenth century, beginning with the 1802 Health and Morals of Apprentices Act. This was not just liberal humanitarianism, it was an inevitable tendency of industrial processes to become more and more self-acting and to require skilled labour than children could not provide.

Large working-class families were probably already incompatible with working-class domestic economy when the Education Act of 1876 made education up to age ten years compulsory, at first with no provision for state payment of the modest fees. This and subsequent extensions ensured that wealth flowed generally from parents to children with little chance (except perhaps in agriculture) for any return, whatever the case may have been previously (see Caldwell 1982). Older children might still contribute for a while to the household economy, but the long-standing tradition of adolescents leaving home to work as farm servants in other households had always minimized their contribution to household economy; indeed it can be regarded as a device to minimize the cost of under-employed adolescents (Kussmaul 1981). It may be possible to construct plausible *post-hoc* material explanations for the social class differentials in fertility decline. But the general economic rationality of lower fertility seems apparent at an earlier, possibly much earlier, date. Changes in attitudes and ideology in a modernizing and more secular society (Banks 1981) are emerging as more subtle explanations more compatible with the international evidence from contemporary, less industrialized countries (Simons 1986, Cleland and Wilson 1987).

TWENTIETH CENTURY FERTILITY DIFFERENTIALS

By the end of the nineteenth century these social differences in fertility had become obvious. They caused concern because of their implications for the social composition of future generations, the future of national intelligence, and because the decline seemed to herald a 'twilight of parenthood'— a new era of low fertility and uncertain population growth throughout society. These worries prompted the novel enquiries on fertility in the 1911 census—which was also concerned with social differences in infant

TABLE 3.1. Great Britain: number of live births per women in separate social status categories. Ten-year and five-year averages (Group A women: marriages under 45 years of Age Only).

	Social Status Categories									
	1	2	3	4	5 Non-manual wage earners	6 Manual wage earners	7 Farmers and farm managers	8 Agricultural workers	9 Labourers	All status groups
Date of marriage	Professional	Employers	Own account	Salaried employees						
(A) Absolute values										
1890–99	2.80	3.28	3.70	3.04	3.53	4.85	4.30	4.71	5.11	4.34
1900–09	2.33	2.64	2.96	2.37	2.89	3.96	3.50	3.88	4.45	3.53
1910–14	2.07	2.27	2.42	2.03	2.44	3.35	2.88	3.22	4.01	2.98
1915–19	1.85	1.97	2.11	1.80	2.17	2.92	2.55	2.79	3.56	2.61
1920–24	1.75	1.84	1.95	1.65	1.97	2.70	2.31	2.71	3.35	2.42
(B) Ratios: all status groups = 100 (for each cohort)										
1890–99	65	76	85	70	81	112	99	109	118	100
1900–09	66	75	84	67	82	112	99	110	126	100
1910–14	69	76	81	68	82	112	97	108	135	100
1915–19	71	75	81	69	83	112	98	107	136	100
1920–24	72	76	81	68	81	112	95	112	138	100
(C) Ratios: 1900–09 Cohort for each status group = 100										
1890–99	120	124	125	128	122	122	123	121	115	123
1900–09	100	100	100	100	100	100	100	100	100	100
1910–14	89	86	82	86	84	85	82	83	90	84
1915–19	79	75	71	76	75	74	73	72	80	74
1920–24	75	70	66	70	68	68	66	70	75	69

Source: Glass and Grebenik 1954, Part I, Table 41.

mortality—for which an eight-fold social class scale was first introduced (Stevenson 1920) from which the Registrar General's social class scale was developed (Szreter 1985). The more comprehensive 1946 family census (Glass and Grebenik 1954) with its first-ever official enquiry into family limitation (Lewis-Faning 1949) also concentrated on social class questions.

Their results created the classical picture of the linear, inverse relationship between fertility and social class (Table 3.1). It was a characteristic feature of all industrial societies in the first half of this century, from the United States (Wrong 1967) to the Soviet Union (Jones and Grupp 1987). Lewis-Faning's enquiry showed, not surprisingly, earlier adoption of family planning by middle-class women (Table 3.2), its adoption earlier in marriage and after

TABLE 3.2.   Percentage of women in the different social classes using any form of birth control at some time during their married life; England and Wales 1910–1947.

Date of marriage	Social Class		
	I	II	III
Before 1910	26	18	4
1910–19	60	39	33
1920–24	56	60	54
1925–29	58	60	63
1930–34	64	62	63
1935–39	73	68	54
1940–47	67	53	47

Source: Lewis-Faning 1949, Table 7.

TABLE 3.3.   Birth controllers: percentages using appliance methods; by social class and date of marriage

Date of marriage	Social class I	Social class II	Social class III	All social classes
Before 1910	—*	6	—*	16
1910–19	24	27	14	23
1920–24	46	29	28	31
1925–29	64	35	24	36
1930–34	63	45	39	47
1935–39	72	50	47	56
1940 +	77	51	39	57
Total	66	43	34	46
N	(432)	(949)	(410)	(1791)

*less than 10 cases.
Source: Lewis-Faning 1949, Table 40.

fewer children and an earlier concentration on more effective (appliance) methods as opposed to coitus interruptus (Table 3.3). Unplanned children—initially widespread throughout society—became more concentrated in manual worker's families in later marriage cohorts. Rather surprisingly, and possibly under the pressure of the depression, by the 1930s there was little social difference in overall contraceptive usage between Lewis-Faning's three social classes in the late 1920s and early 1930s, although differences later reasserted themselves (Table 3.2).

SOCIAL CLASS AND FAMILY PLANNING TODAY

By the marriage cohorts of the 1960s, there was little difference in the 'ever-use' of family planning between women from 'manual' and 'non-manual' backgrounds (Glass 1971). None the less, big differences in the efficacy of family-planning methods most popular in different social classes lasted well beyond the time of the Second World War. In the late 1960s for example, 56 per cent of managerial and 52 per cent of other non-manual workers' wives used reliable methods of contraception, while only 45 per cent of skilled manual worker's wives and 38 per cent of other manual workers' wives did so (Woolfe 1971). By 1975 the most effective method of all, the contraceptive pill, was used more by working-class wives (47 per cent of skilled manual workers, 41 per cent of unskilled workers) than by the wives of non-manual workers (32 per cent of professionals, 44 per cent of skilled non-manual workers).

Use of the pill among wives of unskilled workers rose from just 13 per cent in 1967–68 to 43 per cent in 1973, while the proportion of women in professional families using it remained steady at 32 per cent and 31 per cent respectively (Cartwright 1978). However, by 1975 many (45 per cent) women in social class 1 had given it up, compared to 39 per cent n class 2 and 28 per cent of the others. By the 1980s women in the professional and managerial socio-economic groups were just as likely to be protected by the condom as by the pill, and in these regards their contraceptive practice differs strikingly from other groups (Table 3.4). Educated women are more aware of the risks to pill users from pulmonary thrombosis and more inclined to take them seriously. This anxiety, reinforced by more recent worries about cancer, caused a general decline in pill use from 1977. But even in 1975 (Woolfe and Pegden 1976) the sheath was used more, not less, by wives with husbands in the professional and intermediate class (34 per cent and 32 per cent respectively) compared to skilled and unskilled manual (25 per cent and 21 per cent respectively). The cap (diaphragm) is almost exclusively a middle-class contraceptive and never important even among that class, despite the prominence given to the method by the clinics; withdrawal is more a working-class practice. The IUD

TABLE 3.4.   Contraception: current contemporary use by socio-economic group; Great Britain 1983. All Women aged 18–44.

Current methods of contraception usually used	1 Professional	2 Employer/ manager	3 Intermediate non-manual	4 Junior non-manual	5 Skilled/ own account	6 Semi-skilled/ personal service	7 Unskilled	Total
Pill	21	21	35	36	25	33	22	28
IUD	5	8	5	5	7	6	8	6
Condom	21	19	14	8	14	9	10	13
Cap	6	3	3	1	0	1	0	2
Withdrawal	2	3	4	3	6	4	5	5
Safe Period	2	2	2	1	1	1	1	1
Go without, only method	1	1	0	4	1	2	[-]	2
Go without, others	1	0	0	1	0	1	[-]	0
Foam, etc	1	1	0	0	0	0	[-]	0
Other	0	0	0	0	0	0	[-]	0
Total women*	7	7	7	9	9	8	5	8
Total women†	58	55	60	56	54	54	44	55
Others	43	45	40	44	46	46	56	45
Total women (FP status known)	100	100	100	100	100	100	100	100

*At least one method, but excluding first four methods.
†At least one method.
Note: 0 = less than 1%; [–] = 0.
Source: General Household Survey 1983 (unpublished Table FP 8B, OPCS).

(intra-uterine device) is equally popular in all classes, the safe period is little used by any social class (Table 3.4) but more by higher-level non-manual workers than anyone else.

Male or female sterilization are common and increasingly popular: in 1984 over a quarter of all married or cohabiting women aged 18–44 were in unions protected by sterilization (one or both partners); over a third of couples in their late 30s were so protected. In Britain, as elsewhere, sterilization is becoming the most important means of family limitation. We lack any recent detailed family-planning enquiry in Britain. But questions in the annual General Household Survey (e.g. 1986) not only show the extraordinary popularity of sterilization—shared with most developed societies—but also the fact that it is more popular among manual than non-manual families (Table 3.5). Female sterilization is the dominant method in manual workers, male sterilization in the middle class. This terminal contraception is likely to prevent many unwanted last higher-parity births. Determination to avoid pregnancy is also shown in seeking an abortion. Not surprisingly, relevant data relating to social class are very sparse. But statistical data from the British Pregnancy Advisory Service (pers. comm.) and from local surveys (Clarke *et al.* 1983) suggest that women in non-manual occupations are over-represented among those seeking legal abortion, even though they may experience a lower rate of accidental pregnancy than others.

Despite this armoury of contraception, unwanted births, at least in respect of timing, are still common. In the 1970s, 21 per cent and 18 per cent in social classes 1 and 2 said their last pregnancy was unintended and from 27 per cent to 47 per cent of the wives of manual workers from class IIIM to class V

TABLE 3.5. Married or cohabiting women aged 18–44: percentage of women and partners sterilized for contraceptive reasons, by social status of partner and age of woman.

Age of woman	Social status of partner		
	Non-manual	Manual	Total
18–29			
Woman sterilized	1 ⎫ 4	4 ⎫ 8	3 ⎫ 7
Partner sterilized	3 ⎭	5 ⎭	4 ⎭
30–44			
Woman sterilized	14 ⎫ 35	20 ⎫ 39	18 ⎫ 37
Partner sterilized	21 ⎭	19 ⎭	20 ⎭
Total			
Woman sterilized	10 ⎫ 25	14 ⎫ 28	12 ⎫ 27
Partner sterilized	15 ⎭	14 ⎭	14 ⎭

Source: General Household Survey 1984, Table 4.23.

(Dunnell 1979). But this is a substantial improvement over the position in the inter-war years reported by Lewis-Faning (1949), where over 40 per cent of births in most classes were unwanted. Education is obviously important in these differences. Educated women have fewer unwanted babies; they are more likely to use family-planning methods—especially the more effective ones (Woolfe 1971). The diffusion of reliable methods has become so general that there is now little difference by status. In fact by the 1970s, fewer university-educated women took the contraceptive pill—just 21 per cent compared to 43 per cent of other women. Of university educated women 48 per cent used the sheath compared to 27 per cent of the rest, for reasons discussed above.

THE PRESENT PATTERN OF FERTILITY BY SOCIAL CLASS

The censuses of 1951, 1961, and 1971 and the fertility census of 1946 show how the social class patterns of fertility evolved since the mid-century. More recently, census data have been reinforced by new analyses from the registration of births since 1970, the General Household Survey since 1971, and other surveys. On the whole these show that the relationship of fertility to class has changed from a straight line to a J-shape, then, at least at some ages of

TABLE 3.6. Married women married once only at ages under 45 and enumerated with husband. Mean family size at 1971 census by social class of husband and year of marriage (10% sample), England and Wales.

Period of marriage	Husband's social class						
	I	II	III (NM)	III (M)	IV	V	Total
1941–45	2.04	1.99	1.86	2.20	2.24	2.47	2.14
1946–50	2.11	2.02	1.90	2.24	2.29	2.57	2.19
1951–55	2.25	2.17	2.00	2.34	2.36	2.66	2.29
1956–60	2.23	2.12	2.00	2.29	2.31	2.58	2.25
1961–65	1.80	1.76	1.66	1.89	1.93	2.14	1.85
1966–70	0.60	0.66	0.62	0.85	0.92	1.05	0.79
All periods	1.75	1.81	1.62	1.95	2.03	2.26	1.68
All periods, by woman's class	1.45	1.61	1.35	1.65	1.95	2.39	1.70

NB Fertility of recent marriage cohorts, especially 1966–70, is incomplete at 1971. 'Inadequately described' and 'others' omitted from table.
Sources: Fertility report from the 1971 census OPCS Series DS no. 5 Table 5.7, Table 5.30. Census 1971 fertility tables, vol. II, table 24.

women, a U-shape. More recently the differences have become difficult to discern at all with simple measures (Werner 1985).

Since the 1950s or even earlier the lowest fertility is to be found among families where the husband belongs to social class III(NM) (routine clerical workers, clerks, salesmen). On each side of this central social class, fertility then rises. This has been apparent in the completed fertility of marriages since the Second World War (Table 3.6). Even in the census of 1951 average family size of married women age 25–29 years with husbands in social class I was 1.57 children compared to 1.48 for social class 2. By 1961 social class II fertility had risen to be about as high as that of class III, while that of IV and V—semi-skilled and unskilled workers—had fallen by a tenth or more. By 1971 the trend had developed from a roughly linear inverse association of class and fertility through a J-shaped curve, with fertility lowest in social class III(NM) and highest still in social class V.

Since the middle 1970s it has been possible to chart social class trends in fertility rates with more precision without depending on the census, because the Labour Force Survey gives biennial (annual since 1984) estimates of the population by social class to act as the denominator for the calculation of rates. Up to 1977 fertility in social class I moved even closer to that of unskilled manual workers, as fertility of the latter fell faster (Table 3.7), although this trend has now halted (Werner 1985).

Although the averages of fertility of social classes I and II are closer now to that of class V the patterns of family formation remain different. Fewer women in social class I or II remain childless or have just one child than in social class IV and V (Table 3.8). In social classes I and II the distribution of family size is more concentrated on two- and three-child families. Families with more than three children are more frequent in the manual social classes, especially class V.

None the less trends analysed by birth order and by age of mother reveal a relative increase of fertility in classes I and II. While maintaining or slightly increasing their relative fertility in respect of first and second births, their position in respect of third and even fourth births has declined less than other social classes and the birth rates are correspondingly closer. There are even sharper trends by age. All classes have shared a substantial decline in legitimate teenage births. But while birth rates to social classes I and II women aged 20–24 have declined the least of any class from 1970 to 1983, these two non-manual social classes have increased their birth rates for first legitimate births from 25–29 (though not as much as class III (NM) to V). Their fertility over age 30 is now clearly the highest of any class—a consequence of the remarkable delay in childbearing in non-manual families.

The fertility of social classes I and II contributes less to absolute numbers of births per year than these figures might suggest. Marriage is later and first

TABLE 3.7.   Age-specific legitimate birth rates* by social class of father,
1970–1983, England and Wales.

Age of mother	Year of birth	All social classes	I and II	III (NM)	III (M)	IV and V	Others
		Legitimate births per 1,000 married women by social class of father†					
All ages	1970	114	95	110	117	131	111
	1973	96	89	93	96	105	104
	1977	81	80	79	79	90	68
	1980	92	86	93	91	107	71
	1983	87	82	91	84	101	77
Under 20	1970	465	356	362	452	545	430
	1973	353	283	243	357	410	287
	1977	305	209	207	315	371	198
	1980	347	265	251	356	446	152
	1983	327	290	302	327	386	203
20–24	1970	246	189	203	258	307	204
	1973	206	177	163	216	236	191
	1977	171	142	128	182	209	131
	1980	206	165	167	217	258	145
	1983	207	170	166	212	254	180
25–29	1970	174	191	188	165	173	128
	1973	154	183	166	143	136	133
	1977	139	170	147	126	125	98
	1980	162	181	174	157	147	108
	1983	163	173	174	157	160	119
30 and over	1970	43	44	43	43	45	30
	1973	34	38	34	31	34	29
	1977	32	42	35	26	27	20
	1980	39	50	44	32	33	25
	1983	38	49	44	30	32	24

* The rates for women of all ages, under 20 and 30 and over are based on populations of married women aged 15–44, 15–19 and 30–44 respectively.
† The rates for 1970, 1973 and 1977 are based on the 1970 Classification of Occupations. Rates for 1980 and 1983 are based on the 1980 Classification of Occupations.
Source: Werner (1985), Table 5.

birth intervals (between marriage and first birth) are longer, so it takes them longer than manual workers to produce their completed family sizes. Hence their period aggregate fertility rates are correspondingly lower. Generation length (age at median birth) in social classes I and II increased from 28.1 years

TABLE 3.8. Family-size distribution by social class at ten years duration of marriage; women married once only, enumerated with husband.

Social class of husband	No. of children						N of families
	0	1	2	3	4	5 +	
I	9	12	46	26	6	2	1659
II	10	15	43	23	7	1	4989
III (NM)	13	19	42	20	5	1	2561
III (M)	9	15	42	23	8	3	10 065
IV	11	16	37	21	10	4	3698
V	12	13	29	23	14	9	1243
Social class of wife							
I	17	17	36	22	[—]	[—]	105
II	17	16	44	18	4	1	1874
III (NM)	24	21	39	13	3	[—]	3210
III (M)	20	22	36	17	1	[—]	850
IV	14	17	43	18	6	2	2975
V	8	16	43	22	8	2	891

NB: [—] = less than ten cases.
Source: Censuses 1971, England and Wales fertility tables, volume II, Table 38, Table 40 (10 per cent sample). HMSO, London.

in 1970 to 29.5 years in 1983, compared to 25.7 and 26.0 respectively for classes IV and V (Table 3.9).

Furthermore, women in social classes I and II have a smaller share of the growing number of births which are illegitimate, either unplanned or within a non-marital union (many involving a divorced woman). Social class data from birth registration are only available when the births are jointly registered. Ten per cent of all births in 1983 were jointly registered in this way; but only 5 per cent of classes I and II births compared with 16 per cent of classes IV and V births (Table 3.10). The Longitudinal Study (see Werner 1984) can show the social class origin of all illegitimate births by linkage to the household of origin and thereby to the social class of its head; often the mother's father (Fig. 3.1). This shows a five-fold difference between classes I and V in the proportion of girls having an illegitimate birth before their twentieth birthday, for example. And social classes I and II have a smaller share of the increase in births to remarried women, now about 10 per cent of all births. This is because divorce and remarriage are more common among manual workers.

Higher fertility in classes I and II therefore is due to relatively few families having no children or only one child, a high proportion having just two or three children, and only a few families having more than three children. Manual workers have higher than average fertility because more of them have

TABLE 3.9. Mean ages of women at childbirth within marriage according to social class of husband.

	Mean ages (in years) at legitimate births											
	All births			First births			Second births			Third births		
	1975	1984	1985	1975	1984	1985	1975	1984	1985	1975	1984	1985
All classes (including 'other')	26.6	27.6	27.8	24.7	25.8	26.0	26.7	27.8	27.9	28.8	29.7	29.8
I and II	28.2	29.6	29.7	26.7	27.8	28.0	28.5	29.9	29.8	30.5	31.8	32.1
III (NM)	27.2	27.9	28.1	25.6	26.4	26.5	27.6	28.4	28.5	29.9	30.3	30.6
III (M)	26.1	27.0	27.1	24.0	25.1	25.2	25.9	27.0	27.1	28.4	29.0	29.0
IV and V	25.4	26.1	26.1	23.0	23.8	24.0	25.2	25.7	25.9	27.3	27.8	27.7

Source: Fertility Trends in England and Wales 1975-85, OPCS Monitor FM1 86/2.

TABLE 3.10. Distribution of births conceived inside and outside marriage by age of
mother and social class of father 1973 and 1983, England and Wales.

	Percent legitimate births conceived before marriage		Illegitimate births jointly registered by both parents*	
	1973	1983	1973	1983
All social classes	8.9	6.8	4.2	10.3
I and II	4.5	4.5	2.5	5.0
III (NM)	6.3	5.4	2.4	6.1
III (M)	9.7	7.5	4.7	11.9
IV and V	12.7	8.8	6.1	16.3

*As percentage of all births registered with details of father.
NB: This distribution does not include illegitimate births which were not jointly registered.
Source: Werner (1985), Table 8.

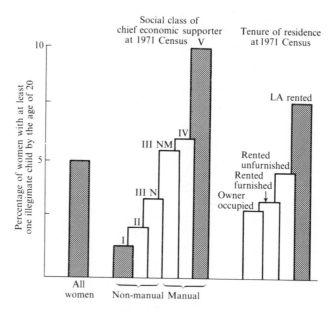

Fig. 3.1. Percentages of women born 1955–59 who had an illegitimate child before their
twentieth birthday. Source: Werner (1985), figure 4.

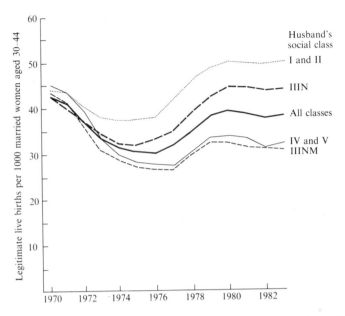

Fig. 3.2. Legitimate fertility rates for married women aged 30 and over by social class of husband, 1970–1983, England and Wales. Source: Werner (1985).

big families, not because they avoid having really small ones. Of the wives of manual workers married in 1955–60, 41 per cent have at least 4 children compared to 29 per cent in the non-manual group. Earlier marriage is the main reason, so that childbearing starts sooner and consequently lasts longer.

Women with husbands in social class I keep their fertility tightly rationed even when they do marry early. Social class I women married under age 20 have an average 1.96 children, declining to 1.23 for women married at age 30–44. The gradient by age at marriage in other classes is more steep; more like a natural fertility decline. In class I I I(M) the equivalent decline by age is 2.33 to 1.00, and 2.69 to 1.11 in class V. There are no significant social class differences in family size among women married at age 25–29. Women in social class I married over age 30 have higher birth rates than anyone else (Fig. 3.2).

SOCIAL DIFFERENCES IN AGE AT MARRIAGE AND TIMING OF BIRTHS

Typically, as well as marrying 2–3 years later, middle-class couples start their families two years later after marriage. From the 1930s until the 1970s the

TABLE 3.11. Legitimate birth by social class of father, England and Wales 1970–1985.

		Median interval from first marriage to first birth (months)			
	All social	Social class of father			
Year of birth	classes	I and II	IIIN	IIIM	IV and V
1970	19	29	26	19	13
1973	24	33	30	22	15
1975	28	37	35	25	18
1977	29	40	36	27	19
1980	29	41	37	25	18
1983	29	39	34	27	20
1985	28	37	33	26	18

Source: Werner (1985), Table 7. OPCS Monitor FM1 86/2, Table 11.

average couple's first baby arrived about 20 months after their marriage. But during the 1970s the gap widened to 30 months—a new pattern. Brides with husbands in social class I had their first baby on average 32 months after marriage in 1971, but 44 months in 1979—a full year later, an increase of 38 per cent. But during the 1980s data from the Longitudinal Study shows that these very long first birth intervals in classes I and II have shortened somewhat (Table 3.11) although the continued increase in age at marriage has kept mean age at maternity high, at about 29 years (Werner 1988).

In 1975, women with husbands in social class V gave birth for the first time on average a mere 9 months after marriage, as at most times in the past. A high proportion (47 per cent in 1971) were premaritally conceived. Now this interval has increased by a half (49 per cent) to 13 months. Much of this shift is due to the reduction in premarital conceptions, which are a major precipitating factor in marriage (Britton 1980) especially in social class V. If premarital conceptions are eliminated from the calculation, the social class gradient in the duration of time from marriage to first birth is only half as steep as before, from 45 months in social class I to 24 months in class V. In this respect, as in many others, there seems to be a sharp difference between the behaviour of social class V and other groups.

BIRTHS AND THE SOCIAL ORIGINS OF THE NEXT GENERATION

The 1911 census contributed to contemporary fears that the better educated classes were less represented in each generation than the less educated.

The elementary methods of ability measurement then available (like their more refined successors today) showed an association between higher social class and intelligence. Half a century of national debate followed on the supposed downward trend of average national intelligence. It was given particular attention by the Royal Commission on Population (Royal Commission on Population 1950). But post-war measurements indicated a slight rise, not the predicted fall, in national intelligence, despite the persistence of the negative association of family size with test score (Thomson 1946). Since then the subject has been muddied by the exposure of the fraudulent data of Sir Cyril Burt. But irrespective of its cause, the lowest average measured ability is found in the social classes with the highest average fertility, although there is considerable overlap in both variables (see Mascie-Taylor, this volume). The topic has always been controversial (see e.g. Searle 1981). It is nowadays difficult to

TABLE 3.12. Estimated legitimate live births by social class* of father, 1971–85, England and Wales.

Year	Thousands of births					
	I and II	III (NM)	III (M)	IV and V	Other	All
1971	154.7	75.3	297.9	160.1	29.5	717.5
1976	140.8	55.5	204.6	110.2	19.4	530.5
1981	163.3	60.4	198.2	111.4	20.1	553.5
1985	158.4	56.2	184.3	105.9	25.4	530.2

Figures expressed as a percentage of all births in each year (omitting 'other')

Year	I and II	III (NM)	III (M)	IV and V	All
1971	22	11	43	23	100
1976	28	11	40	22	100
1981	31	11	37	21	100
1985	31	11	37	21	100

*Broadly speaking the social class categories are:
     I Professional occupations
    II Intermediate occupations
III (NM) Skilled occupations (non-manual)
  III (M) Skilled occupations (manual)
    IV Partly-skilled occupations
     V Unskilled occupations
  'Other' Includes armed forces and students
Note:
(i) The 1981 and 1985 figures are based on different occupation definitions and thus are not strictly comparable with the earlier figures.
(ii) Percentages may not cast to 100 because of rounding.
Source: OPCS Birth Statistics Series FM1 No. 4 1977, t.11.1. OPCS Birth Statistics Series FM1 No. 12 1985, t.11.1.

refer to it in public. Sir Keith Joseph, when Secretary of State for the Social Services, revived the controversy in 1972 to his political detriment by attributing the persistence of some social problems to the unplanned high fertility of girls from unskilled manual backgrounds (Madge, 1988).

Births in general fell by a quarter between 1970 and 1975. But the different social classes shared most unequally in the fall. The number of births to women with husbands in classes I and II remained roughly constant; births to routine non-manual workers (IIIN) fell by roughly the national average, while births to women in social classes IV and V fell by a third (Table 3.12). Consequently, in 1971 about 22 per cent of legitimate births occurred to families in social classes I or II; by 1982 this had risen to 32 per cent, about the same as it was in 1985.

This startling shift in the distribution of births between classes has now stopped at a new balance. But it had already made an important change in the class origins of future generations. In 1971, for example, there were about 224 000 births to fathers in non-manual groups, compared to 469 000 to 'manual' fathers—a ratio of 1 : 2.1. By 1980, 'non-manual' births were almost

TABLE 3.13.   Distribution of male population aged 15–64 by social class at time of census 1911–1981.

(a) Uncorrected

		Percentages						
		1911	1921	1931	1951	1961	1971	1981
I	Professional, etc. occupations	11.2	2.3	2.4	3.3	3.8	5.2	5.4
II	Intermediate occupations	16.1	18.9	13.2	14.5	15.2	18.1	22.4
III	Skilled occupations	34.5	42.6	48.7	52.9	51.0	50.9	47.8
IV	Partly skilled occupations	22.6	22.0	18.1	16.1	20.8	17.7	17.7
V	Unskilled occupations	15.8	14.1	17.5	13.1	9.1	8.0	6.7

(b) Corrected

		Percentages						
		1921 on 1951	1931 on 1951	1951	1951 on 1981	1961 on 1981	1971 on 1981	1981
I	Professional, etc. occupations	2.4	2.5	3.3	2.6	3.4	4.8	5.4
II	Intermediate occupations	15.3	13.8	14.5	13.8	15.8	19.2	22.4
III	Skilled occupations	46.9	49.4	52.9	52.1	52.1	50.7	47.8
IV	Partly skilled occupations	21.7	18.4	16.1	22.6	19.1	17.6	17.7
V	Unskilled occupations	13.8	15.7	13.1	9.8	9.3	7.6	6.7

Source: Boston (1984).

the same as before, 226 000; 'manual' births had declined to 333 000, a ratio of 1 : 1.6. In 1985 the respective numbers were 215 000 and 290 000, a ratio of 1 : 1.4 (OPCS 1986).

This is not just due to changes in fertility. The whole occupational distribution of the population is becoming more middle class (Werner 1985) and therefore more 'non-manual' in composition as industrial jobs—especially unskilled and semi-skilled—are lost and are replaced with higher-skilled manual jobs and especially non-manual jobs in the service sector, including professional. The trend towards more skilled work in general has been in progress throughout much of the industrial period (Table 3.13). Industrial, and particularly skilled manual work reached its peak in 1951 when Great Britain became the most industrialized nation on earth, before or since (Champion *et al.* 1987). This has powerful implications for patterns of consumption, for political change (Heath and McDonald 1987), and for social policy, for example in university admissions (Royal Society 1983).

CLASS, WORK, AND FAMILY SIZE

Nowadays most married women go out to work. This is one of the most important changes in our society this century (Ermisch 1983; Martin and Roberts 1984) and it has important implications for the demography of social class. In 1931 only 10 per cent of married women were in paid employment. By 1966 this figure had already risen to 38 per cent. In 1982 61 per cent of married women aged 16–54 were economically active; that is, at work or looking for a job (this includes women in part-time occupations, which have increased disproportionately). Wives of manual workers are still more likely to go out to work: 53 per cent of wives of men in class I I I (N M) in 1981, compared to 48 per cent of wives in class V and 41 per cent in class I. But the most dramatic increase has been in the proportion of wives in middle-class families going out to work (Table 3.14) and returning to work after the birth of their last child (Joshi 1985).

Economist and demographers are excited by this trend, because women in work generally have fewer babies than women married at the same age who remain housewives (Jones 1982). Mean family size in 1971 of women married in 1951–55 was 2.06 for employed women, 2.61 for housewives—higher by 27 per cent (OPCS 1983). More employed women are childless, more 'housewives' have large families (over 4 children) and the difference is growing. Fertility of housewives has been less volatile than that of women with jobs: it neither increased so fast in the 1950s nor fell so fast since. In fact the early stages of such a movement into the workforce might be expected to increase the birth rate because it encourages women to compares their childbearing into a shorter span of their married life and therefore to reduce birth

TABLE 3.14. Workforce participation rates: Great Britain 1911–1981.

	1911	1921	1931	1951	1961	1971	1981
				All females			
All ages	32.3	32.3	34.2	34.7	37.4	43.6	45.5
<20	38.8	48.4	70.5	78.9	71.1	56.0	59.5
20–24	61.9	62.4	65.1	65.4	62.0	59.4	68.3
25–44	29.3	28.4	30.9	36.1	40.8	51.4	59.5
45–64	21.6	20.1	19.6	28.7	37.1	50.1	51.9
>65	11.5	10.0	8.2	5.3	5.4	6.8	3.7
			Single, widowed divorced females				
All ages	—	53.8	60.2	55.0	50.6	44.7	42.9
<20	38.9	48.8	72.1	80.7	73.2	62.8	76.1
20–24	77.6	80.5	84.0	91.0	89.4	81.7	81.8
25–44	70.3	69.3	74.5	81.2	84.2	80.5	77.0
45–64	44.7	44.3	43.9	50.2	57.4	57.7	52.5
>65	14.4	12.7	10.9	6.6	6.5	6.7	3.4
				Married females			
All ages	9.6	8.7	10.0	21.7	29.7	42.9	47.2
<20	12.6	14.6	18.7	38.1	41.0	43.3	49.5
20–24	12.1	12.6	18.5	36.5	41.3	45.8	54.8
25–44	9.9	9.1	11.5	25.6	33.6	47.4	53.1
45–64	9.3	8.0	7.7	19.0	29.6	47.7	50.7
>65	4.9	4.2	2.9	2.7	3.3	6.9	7.1

Source: Hatton (1986). Centre for Economic Policy Research Discussion Paper 113 *Female labour force participation: the enigma of the interwar period*, Table 1.

intervals (especially the second) (Ni Bhrolchain 1985, 1987). But its more recent effect has been to delay childbearing after marriage (De Cooman *et at.* 1987). Higher fertility of housewives, compared to that of employed women, is greater further down the social scale (OPCS 1983).

Throughout the industrial world, the post-war movement of married women into the workforce is thought to be the main reason why fertility has fallen so low and been delayed so late in the last 15 years (Ermisch 1979; Davis *et al.* 1986; De Cooman *et al.* 1987). This is because childbearing must now compete with new opportunities for married women to work, which puts a substantial price on children and especially on larger families and long birth intervals.

In the USA, Espenshade (1984) estimated that the direct costs of a child raised to age 18 ranged through $75 000 for low socio-economic status husband to $98 000 for high-status husbands (1981 dollars) and in addition to that a further four years at public college would cost $15 000 and four years at a private college $27 500. Later estimates (Calhoun and Espenshade 1988) suggest that the total cost of each child in a typical middle-class American family with two children is about $104 000 (1981 dollars) from birth to age 18, of which direct costs comprise 80 per cent. British estimates point to a total lifetime earnings loss for a two-child family from £113 000 to £145 000 spaced from two to six years apart (Joshi 1987) and that rearing a two-child family takes seven years of a woman's working life, and 13 years' equivalent income if the loss of promotion is taken into account.

Furthermore, especially since the Equal Pay Act of 1970, women's average hourly earnings have risen from about 55 per cent of men's to about 75 per cent (full-time work, without overtime). This has benefited working-class women and correspondingly increased the opportunity costs of their child-bearing. Women barristers, civil servants, teachers, and scientists, on the other hand, have always been paid the same as men for employment at the same level (although woman teachers and civil servants were required to resign upon marriage until the Second World War). Women in business (now 8 per cent of middle management) are catching up in the promotion structure.

The size of these costs suggest that they are likely to be the driving force behind emergent patterns of social class differential fertility and patterns of child spacing. Both for men and for women, occupations which determine higher social class membership are also better paid, so the opportunity cost of childbearing will rise in relation to women's occupation. The relation of earnings to age also differs by social class. Many manual jobs have no career structure and no increments related to age or experience. But in many professional and business jobs, the need to get established and the higher costs of being away from work put an economic premium on deferred childbearing, short birth intervals, and moderate family size.

Discounting the cost at present value, to give a more realistic estimate of the loss at the time decisions are made, emphasizes the importance of deferring births. In Joshi's (1987) analysis, producing two children early cost £84 000, having them later costs £62 000 other things being equal. On the same basis, a six-year interval costs £89 000 compared with £72 000 for a two-year interval. Such analysis has not yet been extended to the particular question of social class differences in fertility or marriage. Wives with jobs have a social class in their own right which has a bigger effect on their fertility than the social class of their husbands. The 1971 census (OPCS 1979, 1983) showed that wives who are themselves in professional jobs (social class I) have lower fertility than wives with jobs in social class II, whereas the wives of men in social class I have higher fertility than the wives of men in social class II. Perhaps it is more astonishing that professional women should have any children at all—certainly a high proportion choose to remain unmarried (Table 3.15), and their rates of illegitimate birth are also particularly low.

Part of the answer may lie in child-care support for working mothers.

TABLE 3.15.   Number of children expected in all by year of birth of women and
socio-economic group of woman's father.

Year of birth of woman	Socio-economic group of woman's father	Average no. of children expected
1935–39	Non-manual	2.22
	Manual	2.53
	Total	2.47
1940–44	Non-manual	2.18
	Manual	2.40
	Total	2.37
1945–49	Non-manual	2.02
	Manual	2.22
	Total	2.17
1950–54	Non-manual	2.09
	Manual	2.19
	Total	2.17
1955–59	Non-manual	2.23
	Manual	2.15
	Total	2.18
1960–64	Non-manual	2.30
	Manual	2.25
	Total	2.25

Source: General Household Survey (1983). Table 4.24.

Preschool facilities are mostly used by children of middle-class parents, although in 1980 there were only places in all such institutions for 4 per cent of the 0–4 age group (Osborn et al. 1984). Only day nurseries (and childminders) provide full-time care; these are a particular help for divorced or single working mothers. For a small minority of women in better-paid jobs, domestic service may support fertility. Women with professional or business careers may not be able to sustain both job and children without help from a nanny or au pair. This aspect of contemporary fertility is relatively unexplored (Ermisch 1988). Residential domestic help was regarded as essential for the maintenance of the large Victorian middle-class family, even though the wife never worked. The general absence of domestic help in the modern home is the most striking, possibly the only, real reduction in middle-class living standards since Victorian times, or even the 1920s.

When a household has two incomes, two sets of occupational influences from the outside world impinge on fertility. Then it makes sense to look at the joint social class position of working husbands and wives for their influence on family size (Britten and Heath 1983; Osborn and Morris 1979). The General Household Survey and the 1971 census show that what the wife does for a living determines family size more than the husband's job. Whatever the husband's job, non-manual employment of the wife brings lower fertility, especially when the husband is a manual worker (Registrar General 1979). Similar patterns emerge from the *Child health and education study* (Britten and Heath 1983). Some cross-class marriages have family sizes well away from the average. Where the husband is in social class I or II and the wife in social class III(NM) only 24 per cent have more than two children. Wives in classes I and II married to husbands in manual occupations have much higher fertility—42 per cent have family sizes greater than two. There are, therefore, big differences in family building patterns determined by whether the wife is employed and the standing of her occupation in relation to that of her husband.

FAMILY INTENTIONS

Social class patterns of 'ideal' family size are rather uniform in England and Wales, but 'expected' family size shows larger class differences. For example, in the 1960s, the mean number of children considered ideal for families 'like yourselves' (wives aged 25–29) was 2.4 in all classes except the wives of the unskilled and semi-skilled (2.5). The ideal number for families 'with no particular worries about money' was almost uniformly one child higher. But when couples were asked how many children they were likely to have in reality, bigger social class differences became apparent. On average, managerial respondents expected 2.6; skilled and other workers 2.5.

Since 1945, manual workers' expectations of family size have fallen, managerial workers' expectations had risen. Women born in 1935–39 with 'non-manual' fathers expected 2.10 children compared to 2.52 for women with 'manual' fathers. By 1955–59, women with non-manual fathers expected rather *more* children (2.24 to 2.12) than women from manual backgrounds (Woolfe 1971). More recently these expected family sizes have declined across all classes, but middle-class expectations remain slighly higher (Table 3.15) (General Household Survey 1985).

In France in 1982, ideal family size for 'people in your circumstances' was smaller with lower social status. Professionals, manufacturers, and senior management have by far the highest fertility ideals, more than farmers (who traditionally have large families), with labourers and agricultural workers on

TABLE 3.16.   Ideal family size by social group in France 1982.

Occupation of head of household	Ideal family size (average)	
	In general	In your circumstances
Farmers	2.71	2.46
Agricultural workers	2.60	2.29
Labourers	2.55	2.22
Skilled workers, tradesmen	2.69	2.42
Employees	2.59	2.23
Middle management, technical specialists	2.66	2.49
Higher management, manufacturers, professionals	2.70	2.75

Source: Bastide *et al.* (1982), Annexe II.

average half a child lower (Table 3.16). Other societies are experiencing a convergence in expected fertility, but the cross-sectional data from the US 1980 Current Population Survey do not yet show a higher expected fertility among those in the highest income bracket given (over $24 999) compared to others, although among whites differences are small except for the lowest income bracket (less then $7 500) (Wineberg and McCarthy 1986). But $25 000 is not a high income.

EDUCATION AND FERTILITY

Social class effects are strongly associated with educational differences. The connection may explain part of the fertility differences by class, through knowledge of the needs and methods of family planning and the ability and foresight to plan.

The effects of education on fertility are likely to be increasingly important, as a much higher proportion of both sexes now have higher educational qualifications than formerly. Only 5.4 per cent of women born as late as 1922–26 had qualifications at 'A'-level or its equivalent, little more than women born in 1902–1906 (4 per cent). But this has increased to 10.5 per cent for the cohort born in 1937–41, and to 16.2 per cent for the 1952–56 cohort (1981 census, Qualified Manpower tables).

Unfortunately most data relating education to fertility are rather crude. Furthermore, they are not comparable over time. Earlier censuses (1951, 1961) asked about terminal age of education; more recent ones (1961, 1971, 1981) asked for qualifications. The distribution of both has changed radically. The question on terminal age was not repeated in 1971; most of the population give the minimum age, which itself has increased over time. But since 1961 a question has been asked to list all educational qualifications, which divides the population into two or three very unequal sections on the basis of the possession of 'O'-level qualifications or higher-level qualifications. This does not permit a very subtle analysis.

In general, people with more education have fewer children, in part because they marry later. As might be expected (Table 3.17) women who left school below age 15 had the largest families, but the best educated had the next highest fertility. Furthermore, between 1951 and 1961 the fertility of better educated women increased relative to the less educated, amidst a general decline of family size. The trend according to educational qualifications from 1961 to 1971, shows a similar pattern.

Various cross-sectional surveys in the 1970s (e.g. Cartwright 1978) confirmed that the average family size of graduate women at the time of interview (1.53) was considerably lower than the fertility of women with no further eduction (2.04), although the graduates' deficit is exaggerated in cross-sectional studies by their longer generation length. The picture is the same elsewhere, for example in the USA where the broad educational categories used show a more or less linear, declining relationship with women's education. Unfortunately, no data were given on fertility by men's occupation (Wineberg and McCarthy 1986).

TABLE 3.17. Education and fertility; median age at birth of first child.

Qualification	Men	Women
None	24.2	20.9
'O'-level*	24.7	23.3
'A'-level*	25.9	25.1
Advanced (non-degree)	27.3	25.8
Degree	28.1	28.1
School level age		
15	24.6	21.6
16	25.7	24.2
17	27.0	25.3
18 or over	28.1	25.9
Ability scores at age 11 years		
1. low	24.6	22.5
2. median	25.8	23.8
3. high	26.8	25.3

*or equivalent.
Source: Kiernan and Diamond (1982), Table 2, p. 15.

Age at marriage causes most of the difference, although the fertility of graduate women is the least influenced by age at marriage of any educational level. At all ages of marriage over 20 years (women married 10–14 years at the 1971 census) graduate fertility is higher than the least educated women—5 per cent higher among graduates at age 20–22½ rising irregularly to 41 per cent higher for graduates married at age 35–39. Graduates are the least likely of any group to have just one child and are also much less likely to have fourth and especially fifth children. In any cross-sectional enquiry, married graduates are much more likely to be childless (25 per cent compared to 24 per cent and 16 per cent in the other two groups) but this is because they marry later and defer childbearing.

Graduate husbands were less likely to be childless than were graduate wives at the 1971 census (OPCS 1979, 1983). This accounts for their higher average fertility. This apparent paradox arises because 52 per cent of graduate wives have graduate husbands, while only 31 per cent of graduate husbands have graduate wives. Graduate wives have the lowest fertility, at any level of husband's education. This is mostly due to childlessness rather than small family size: fertile graduate wives married to graduate husbands have quite high average family size (2.25). Many of these relationships are similar to those noted for social class I. Nowadays most professions require a degree for entry, but most graduates do not become professionals. And as in social class, the effects of wife's education predominate over those of the husband. Differences in expected family size by education have almost disappeared in recent cohorts (Dunnell 1979), another parallel with the shrinking and reversing trend of social class fertility differentials in Britain.

Qualifications, ability, and family history help determine the age at which childbearing starts (Kiernan and Diamond 1982) and therefore final family size. The less able (measured at age 11) and those with no qualifications start having babies earlier. So do women with many brothers and sisters or whose mothers themselves married young. Women still childless at age 32 were more likely to be from a professional background with few brothers or sisters, and with mothers who themselves married late. Women with high personal ambition and parental interest also tended to defer child-bearing. There does, therefore, seem to be some inheritance of patterns of childbearing.

TENURE AND FERTILITY

Social class also has important connections with housing tenure. Indeed for the pioneer 1911 census, it was originally proposed that social class be assigned on the basis of occupation, and of number of rooms occupied, although this idea was later abandoned (Boston 1984). In a recent multivariate analysis of the 1976 *Family formation survey*, housing variables were the most powerful in discriminating between aspects of family formation (Murphy 1987). The effects of housing upon marriage and fertility may help to explain some class differences. Housing tenure in Britain is changing fast, and fertility may change with it.

The fertility difference between council tenants and owner-occupiers is one of the biggest of any between major socially-definable groups (Table 3.18). Tenure brings together many influences under one roof. It combines together the effects of both parents' class, occupation, and attitude. It does not automatically alter with changes of employment. It divides the population into three substantial, although unequal sections: in 1987 65 per cent of households were owner-occupiers, 27 per cent rent from a local authority or a housing

TABLE 3.18. Fertility and housing tenure, Great Britain 1977.

(a) Average number of live births in current marriage to married women aged 40–44 in 1977.

	Housing tenure				
	Owner-occupied	Local authority	Private furnished	Private unfurnished	All
Women in first marriage	2.11	2.95	(1.67)	2.26	2.37
Remarried women	0.75	0.78	(0.50)	(0.13)	0.67
All married women	2.02	2.71	(1.20)	1.95	2.22

(b) Family size distribution of married women aged 40–44

Live births	Owner-occupied %	Local authority %	Private rented %
0	13	10	22
1	17	13	22
2	38	26	20
3	23	22	25
4 or more	9	29	12

(c) Birth interval ever-married women aged 40–43.

Interval marriage–first birth (months)	27	17	24
Average no. of live births	2.29	2.89	2.51
Average age at marriage	23.4	21.7	22.2

Note: ( ) = sample size less than 10.
Source: Murphy and Sullivan (1985), Table 2.3.

association, and 8 per cent rent privately. Apart from the 20 per cent of high-rise blocks, the physical difference between average (purpose-built) local authority housing and owner-occupied housing is small, although the social and environmental differences may be substantial. All social classes are to be found in all tenure groups, although to an unequal degree (Fox and Goldblatt 1982).

The average family size of council tenants is almost one child (0.84) more than owner-occupiers (Murphy and Sullivan 1985). The whole distribution of the fertility of council tenants is different. Childbearing starts earlier: at the 1971 census women in local authority tenure had twice the family size by age 25 of women in owner-occupation. Three times as many had an illegitimate child before their 20th birthday (Werner 1984). Three times as many—almost a third in all—have four children or more. This differs from the usual manual working-class pattern; working-class owner-occupiers are much more like middle-class owner-occupiers in these respects.

The timing of fertility can certainly affect access to housing and therefore tenure. Owner-occupiers often delay marriage and childbearing because they must face heavy housing costs at the beginning of their marriage; council tenants (or intending council tenants) may begin a family quickly because that will increase their chances of being allocated a council house through the points system, even though it reduces their chances of buying or renting privately (Ineichen 1977). Almost half the girls married in their teens and occupying council houses in their mid-20s had conceived their first birth before marriage (47 per cent), compared to 30 per cent owner-occupiers, 14 per cent renting privately, and 9 per cent who were still living with parents (Kiernan 1980). These differences in average family size by tenure persist within each category of educational level and social class (Murphy and Sullivan 1985). Owner-occupiers in different social classes have quite similar achieved family size and intended family size. But among council tenants, family size differs more according to social class (Cartwright 1978). In some local authorities, all available council housing goes to those who have been accepted as statutorily homeless under the Housing (Homeless Persons) Act 1977. In 1986 about 10 per cent of such acceptances were to pregnant single women; this is another route whereby fertility can procure housing and the creation of a new household.

The composition of council tenure is strongly affected by a selective effect, providing housing for those with no alternative: the elderly, unemployed, the less skilled, single-parent families (Department of the Environment 1969, 1977; Murie 1983; Holmans et al. 1987). But cause and effect may also operate in the other direction. The social and physical environment on some of the larger estates is often distinctive (Coleman 1985). The dependence on the municipal landlord for all aspects of housing services and related welfare may be almost complete (Minford *et al.* 1987). Many estates comprise synthetic communities created by relocation (Young and Willmott 1957; Morris and Mogey 1965). It has been claimed that such conditions may in turn influence behaviour, including demographic behaviour. Residents on such estates often have little control over their environment (National Federation of Housing Associations 1985). Some estates are effectively one-class areas with high unemployment, few examples of success, and few alternatives to early marriage or early childbearing for young women. Mobility is particularly low (Salt, in press) because the subsidy in sub-economic council rents is not portable to other tenures, and the points system militates against inward mobility at least by those who are not unemployed, divorced, or pregnant. More council households than average have no earner at all. There are many fewer two-earner households, so the 'housewife effect' on fertility may be important. Clustering of high-fertility families, and geographical and social isolation of some estates, may be important too.

These differences may become sharper in time as housing tenure becomes more polarized. The privately-rented sector, the demography of whose residents is intermediate between owner-occupiers and council renters, is declining fast. Just before the First World War over 90 per cent of households rented privately, by 1939 only 58 per cent were still doing so. Now the figure is just 8 per cent, having fallen from 15 per cent in 1971. Private renting is a transitory tenure for young couples, or a necessity for many divorced or remarried people. Council tenure, non-existent in 1914, grew to a maximum of 34 per cent of households in 1980 and has since declined slightly (see Holmans 1987; Daunton 1987). Government policy since 1979 has vigorously encouraged the sale of council homes to their tenants. Even if only one-fifth more chose to buy, council tenure will be reduced to 22 per cent, and 70 per cent of households would then own their own homes. These trends could make tenants an even more self-selected and marginalized group (Murie 1974).

CLASS, ATTITUDES, AND UNWANTED FERTILITY

Unwanted fertility is an important component in the explanation of class differences in family size. In general, the proportion of births that were unwanted increased from 6 per cent of births to the marriage cohorts up to 1910 to 14 per cent for the marriages of 1930–34. Throughout the first half of this century a much higher proportion of middle-class than of working-class families planned a specific family size (Table 3.19). However, the picture of the disribution of unwanted births is far from clear. Between the wars, middle-class women were most likely to describe births beyond their second as unwanted (Lewis-Faning 1949). Middle-class women were also more likely to

TABLE 3.19. Proportions marrying up to age 40 by social class at marriage rates of
1979 ('gross nuptiality'), England and Wales.

Social class	Males	Females
I	93	86*
II	89	93
III (NM)	84	87
III (M)	86	94
IV	81	86
V	74	98†
All	85	91

*To age 35.
†To age 30.
Source: Haskey (1983), Table 6.

admit that they had attempted an illegal abortion (2.3 per cent of pregnancies in class I compared to 1.0 per cent in class III), but the difficulties of interpretation and reporting are obvious. Surveys in the late 1960s did not show a clear relation between proportion of unwanted births and social class either. Of manual women with at least 4 pregnancies, 31 per cent described their fourth as 'accidental' compared to a corresponding proportion of 24 per cent of non-manual women. But other differences are hard to find (Langford 1976). Part of the problem may lie in not differentiating between those who plan and those who do not, and in the difficulty of asking questions in depth in national surveys. Planned pregnancy rates were consistently lower in the wives of manual workers compared to non-manual workers from 1951–75, except in 1971–75 when planned rates for the non-manual group fell more sharply for first and second pregnancies (Bone 1978).

The position is now somewhat changed. A survey in 1975 (Bone 1978) showed that the wives of manual workers, at a time of declining fertility, had become less prone to unwelcome pregnancy. Sixteen per cent of social class III(M) wives regretted their last pregnancy in the 1975 survey compared to 19 per cent in 1970; 17 per cent of wives in classes IV and V compared to 24 per cent in 1970 (see also Cartwright 1978 and Dunnell 1979). Askham's (1975) study in Aberdeen showed that most of the excess fertility of parents of large families (more than four children) was unwanted, especially in class IIIM as opposed to class V. The parents of small families knew much more about sex, contraception, and family building and had talked much more about them before marriage. The large families in class V were born of ignorance and fear of sex and contraception, and of a reluctance to discuss them. Disapproval of contraception too, and not just ignorance, was a powerful factor. High fertility families were less concerned for the future, had little faith in their

ability to control their own lives, and little ambition. Unfavourable circumstances encouraged strong orientation towards the present, not the future: a
feeling of lack of control over events, and a tendency to accept them passively.

High fertility families in social class V enjoyed much less financial security
than the low fertility families in class III, had less secure and frequently
changing employment, more marital problems and less adequate accommodation. Many were teenage brides, whose problems are well known (Ineichen
1977). The low fertility families in social class III had courted for a long time,
decided carefully about marriage, moved house seldom, and tended to be
upwardly mobile in their jobs.

At the national level, in all classes, families who plan their births have lower
fertility. Their family size clusters more tightly around an average of two
children. They are more likely to use effective methods of contraception.
Considerably more women in the higher social classes have a 'forward-
planning' outlook which assumes control over their futures (Dunnell 1979).
These differences in personality are crucial for control of fertility, and
probably for understanding social class differences.

MARRIAGE AND SOCIAL CLASS

Before the advent of birth control within marriage from the late nineteenth
century, Western European marriage had a unique role in family limitation.
Unlike the rest of the world, individuals could chose to avoid it, or postpone it,
and if they wished, marry whom they chose (Stone 1977; Macfarlane 1986).
Even now with near-universal birth control, early marriage still means larger
families. The earlier marriages typical of the lower social classes thus have a
substantial effect on average social differences in fertility, although it is

TABLE 3.20.   Median age at marriage 1979, England and Wales.

Social class	Bachelors	Spinsters by own class	Spinsters by groom's class
I	25.8	26.0	23.9
II	26.2	24.0	23.2
IIIN	24.5	21.6	22.0
IIIM	25.0	21.2	21.4
IV	23.2	20.4	20.8
V	23.5	22.0	21.0
Armed forces	23.0	—	20.7
All	24.3	21.7	21.7
No. in sample	870	832	832

Source: Haskey (1983), Table 3.

important to remember that marriage is often a public sign that a couple are ready for children, not an independent event which then determines fertility.

Men in social class I are the most likely of all to marry; men in class V the least. The reverse is true among women. Social class V women almost all marry, while women in social class I are particularly likely to remain spinsters. Criteria of marital choice, as well as careers, may matter here. Women tend to 'marry up' and women in class I have nowhere to marry up to. Up to four years separates the average ages at first marriage across the social spectrum (Table 3.20), and the distributions are tightly clustered around their averages. These earlier marriages are just the best recorded aspect of a generally earlier pattern of sexual activity and reproduction among manual workers, especially the unskilled: earlier sexual experience (Schofield 1968; Mant *et al.*, 1988), more frequent cohabitation, higher levels of illegitimate births, more premarital conception (Table 3.11), shorter birth intervals, higher family size, and more frequent divorce and remarriage. However, there seems to be no simple

TABLE 3.21.   Alternative estimates of relative social class marriage breakdown propensity: indexed to SC III M = 100.

Social class	Gibson[1]	Census[2]	Haskey[3]	Murphy[4]	Thornes and Collard[5]
I	76	54	48	70	(6)
II	86	76	86	82	85
III (NM)	148	102	111	97	96
III (M)	100	100	100	100	100
IV	86	118	114	105	102
V	176	182	227	179	(98)
standard deviation	37	40	55	35	—

Notes

(1) England and Wales 1961, derived from Gibson (1974), Tables II and III (also Table 3.1 of this chapter).

(2) Ratio of divorced to married males aged 35–59 in Great Britain in 1971 Census Occupation Tables, Vol IV, Table 29.

(3) England and Wales 1979 derived from Haskey (1984).

(4) Great Britain estimated proportion of first marriages broken by 20 years marriage duration from Family Formation Survey.

(5) West Midlands, 1971 relative risk calculated from sample numbers in Thornes and Collard 1979 Tables 7 and 8, p. 161 according to the formula for i^{th} class; $100*(D(i)/CM(i)/(D(III(M))/CM(III(M))$, where $D(i)$ refers to the numbers in the divorced sample and $CM(i)$ to the numbers in the currently married group for the ith social class. The authors (*op. cit.*, p. 150) reported that 'upper' and 'rougher' elements were definitely more likely to refuse in the divorced sample, and that it was possible that the III M group was over-represented. Results for SC I and V should be discounted.

Source: Murphy (1985), Table 2.

association between social class and (marital) coital frequency between ages 20–45 (James 1974).

Marriage is still the chief setting for fertility, although its pre-eminence is weakening. In the past, up to the 1950s but excluding the war years, about 5 per cent of births were illegitimate, although about 30 per cent of first births were premaritally conceived and saved from illegitimacy by prompt, if not forced, marriage. These events were, and remain, disproportionately common among manual workers (Table 3.10). Thirty per cent of first births to social class V wives were premaritally conceived, compared with 8 per cent in social class II. The Longitudinal Study shows that the chief economic supporter of a teenage girl with an illegitimate baby is disproportionately in class V (Fig. 1). Where the birth is jointly registered the father's class can be assigned. Much of the increase in illegitimacy is due to such births (72 per cent of the total in 1985) and a high proportion have fathers with manual occupations. If there is a movement towards a more Scandinavian system of fertility without marriage, then the unskilled working class is its vanguard.

The circumstances of mariage also exert a profound influence on the chances of subsequent divorce, and therefore on the problems of single-parent families, the demand for more housing, the incidence of homelessness, and the problems of remarriage and step-parenthood. Working-class marriages, especially in class V, are more likely to break down than middle-class marriages, especially those in class I (Table 3.22). This used not to be so. When

TABLE 3.22. Proportion of divorced men and women who remarried within 2.5 years of their divorce, by social class, England and Wales 1979–81.

Social class according to divorced husband's occupation	Husbands %	Wives %	Sample size
I	44	28	25
II	38	23	175
III (NM)	45	34	115
III (M)	33	40	352
IV	28	31	138
V	29	20	69
Non-manual	41	27	315
Manual	31	35	559
Economically inactive	26	35	140
Unemployed	26	28	78
All	34	33	1084

Note: total includes 'armed forces'.
Source: Haskey (1987b), Table 2.

divorce was rare and expensive it was exclusively an upper-class and middle-class practice, although simple abandonment and desertion seem to have been common elsewhere in society, and even the occasional 'wife sale' (Menefee 1981). The experience of two world wars, higher living standards, and the emancipation of women (especially their equal legal treatment in divorce from 1938) opened up divorce, like many other habits, to a wider public. Law reform in the 1940s which made legal aid available for the first time, made divorce practical for the poorest.

Many factors combine to make the marriages of manual workers, especially in class V, more vulnerable than others (Haskey 1984). The circumstances of marriage, the scope of marital choice and the expectations and skills which the partners bring to the marriage are all important. In the unskilled working class, partners are often chosen from a very limited social and geographical horizon (Ineichen 1979; Coleman and Haskey 1986) so despite early sexual activity, experience of alternative partners is limited. Partners chosen from very near home (within 1 km) tend to be divorced early (Haskey 1987a). A partner chosen early may compare unfavourably with others met later. A high proportion of partners in remarriages meet at work (Coleman 1982, 1984), and married women in class V have the highest labour force participation rate. Furthermore, among unskilled workers courtship is shorter and formal engagement less frequent, premarital conception more likely, and a civil ceremony more likely than a religious one. All these are known to be 'risk factors' for subsequent divorce. But the most important one of all is early marriage (Murphy 1985). Earlier marriages are more likely to fail in all classes, and early marriage predominates in the working class.

Working-class marriages may be undergoing a bigger strain of transition than those of the middle class, from the restricted, segregated marital relation-ships so bluntly emphasized in studies of the 1940s and 1950s (Zweig 1961; McGregor 1957; Klein 1965a) to a more rewarding but much more difficult companionate marriage where activities, opinions, decisions, and money are shared (Willmott and Young 1973) and where social skills and the ability to communicate become more important. The abrupt transition described in those earlier studies from the lively independence of a teenage girl to the drabness of a woman's married life will scarcely be tolerable nowadays. Neither will the violence which these studies indicated was—and evidently still is—a particular problem in working-class marriage, especially violence associated with drink. Wives sue for divorce more often than husbands in the working class, and the 'unreasonable behaviour' of their husband is the predominant reason (Haskey 1986). Other studies suggest that this 'unreasonable' behaviour is often violent behaviour. The middle classes apparently prefer adultery but the patterns of petitioning by social class may reflect social norms in those classes rather than actual differences in marital misbehaviour.

One in three of all weddings are now remarriages for one or both partners. Remarriage rates have not gone up—if anything the opposite. The rise in numbers remarrying is entirely due to the increase in the number of remarriageable partners through higher divorce rates (Coleman 1989). Because of class patterns of divorce it follows that remarriage and its attendant problems is slightly more a working-class than a middle-class institution, although the chances of remarriage by class are far from uniform. Divorced men from non-manual backgrounds are more likely to remarry within a given time than men from manual backgrounds—41 per cent within 2.5 years, compared to 31 per cent (Table 3.22). The posiion is reversed for divorced wives—the chance of remarrying for divorced women with husbands from non-manual backgrounds is a third smaller than for women with husbands from manual backgrounds (Haskey 1987b). And the non-manual divorced men were considerably more likely to remarry within two months of divorce than the manual workers (17 per cent compared with 12 per cent); beyond that duration the proportions are more similar. This suggests either that more non-manual divorced men had a new partner already set up (i.e. they may be divorcing in order to remarry), or it may follow from the greater propensity of manual workers to cohabit before their next marriage, perhaps due to more severe material or housing difficulties, or not to remarry at all.

HOUSEHOLD TYPES AND SIZES

These patterns of marriage, divorce, and remarriage create differences in social class patterns of family and household. Lower rates of marriage for lower-class men mean that more of them live alone at all ages. For married couples, higher death rates and divorce rates of manual workers leave more wives living alone too. The predominant importance of divorce in creating one-parent families, combined with the higher divorce risks of working-class women, mean that more children from these origins, especially class V, find themselves victims of a broken marriage and being supported only by their mother (Haskey 1987b). The problems of single-parenthood are suffered disproportionately by working-class mothers and their children (Table 3.23). Divorce affects housing tenure too, being one of the few important channels from owner-occupation to council renting (Holmans et al. 1987). But such broken families may move faster back into a new partnership—formal or otherwise—as the remarriage rates for working-class women are higher than for middle-class women. Furthermore, working-class women may be more likely to cohabit before remarriage (Haskey and Coleman 1986), so increasing social class differences in living patterns still further.

TABLE 3.23. Household type according to socio-economic group of head of household, Great Britain: 1984.

Household type	Socio-economic group of head of household*						
	Professional	Employers and managers	Intermediate and junior non-manual	Skilled manual and own account non-professional	Semi-skilled manual and personal service	Unskilled manual	Total
	%	%	%	%	%	%	%
1 adult aged 16–59	9	7	13	5	9	7	8
2 adults aged 16–59	19	15	15	14	10	7	13
Youngest person aged 0–15	38	37	27	39	27	21	33
3 or more adults	12	13	9	14	11	12	12
2 adults, 1 or both aged 60 or over	19	20	15	17	18	20	18
1 adult aged 60 or over	3	8	21	10	26	33	16
Base = 100%	452	1635	1978	2934	1792	574	9365

*Excluding members of the Armed Forces, full-time students, and those who have never worked.
Source: General Household Survey (1984), Table 3.17.

TABLE 3.24. Proportion of husbands marrying within same social class, by previous marital status.

Social class	First Marriages						Remarriages					
	%	A Base	%	B Base	%	C Base	%	A Base	%	B¹ Base	%	C¹ Base
I	10	49	3	37	27	25	16	25	15	13	14	14
II	43	144	47	101	41	236	31	94	29	55	31	71
III (NM)	68	129	74	95	15	75	62	39	62	26	19	36
III (M)	14	383	11	213	45	364	10	139	12	87	47	139
IV	39	72	21	89	23	144	35	48	38	37	22	68
V	7	43	11	28	18	44	9	11	8	12	20	20
All	29	820	29	563	36	925	25	356	26	230	33	348
tau B	0.33*		0.30*		0.30*		0.24*		0.24*		0.24*	

* Significant at the 0.01 level.
Note: tau B is the Spearman rank order correlation coefficient calculated over the entire contingency table, which is not shown.
Sources
A, A¹   Reading Marriage Survey 1972/73.   Husbands own class, wives own class.
B, B¹   OPCS Sample of Marriages 1979.   Husbands own class, wives own class.
C, C¹   Reading Marriage Survey 1972/73.   Husbands fathers and wives fathers' social class.

MARITAL CHOICE AND CLASS BOUNDARIES

The strength of class boundaries in social life is reflected in the frequency of marriages within and across class lines. Occupation and income may give an objective definition to social class categories and may themselves directly affect behaviour and attitudes (Kohn 1981). But whether social classes continue to have characteristic cultural differences will depend in part on the willingness of young adults to avoid, or to form, friendships with others of different social origins. Even if nominal class membership is constantly recruited through social mobility, no strong cultural differences can be successfully maintained for long against very high rates of in-marriage. They would eventually come to depend primarily on the characteristics required for recruitment to particular occupations, or which the jobs themselves provoked. For these reasons social scientists have long been interested in patterns of social class intermarriage, and for similar reasons intermarriage patterns among religious, ethnic, and racial groups.

Men and women are more likely to marry within their own class (e.g. Berent 1954; Hope 1972), however defined, than into any other. And not surprisingly, the proportions marrying into classes at increasing social distance become progressively smaller. In general, social classes in the middle of society—for example class III(M)—show the highest rates of in-marriage (endogamy) while those at the opposite ends of the social spectrum (I and V) show the lowest. But measuring these differences is troublesome (Coleman 1982). High rates of in-marriage are most likely when the group itself is large; social class III is by far the largest. Women have a markedly different pattern of employment from men, with a high proportion in social classes II and IV,

TABLE 3.25. Ratio of observed marriages to those expected on the assumption of no association (10 per cent sample).

Social class of husband	Social class of wife						Inadequately described occupations
	I	II	IIIN	IIIM	IV	V	
I	2.24	2.06	1.30	0.36	0.34	0.14	0.81
II	2.24	2.23	1.15	0.47	0.52	0.25	0.87
III (NM)	0.70	1.01	1.48	0.70	0.67	0.48	0.81
III (M)	0.23	0.56	0.93	1.33	1.19	1.20	0.97
IV	0.15	0.52	0.74	1.17	1.40	1.57	1.04
V	0.05	0.36	0.52	1.24	1.44	2.34	1.36
Inadequately described occupations	0.61	0.84	0.84	0.63	0.87	1.03	4.03

Source: OPCS Fertility Report from the 1971 Census, Series DS No. 5, Table 5.33.

which tends to increase endogamy rates for men in those classes at the expense of others. It can be more useful to classify brides by their fathers social class. The social isolation of groups can also be shown in terms of the departure from random marriage shown by the actual marriage rates (Table 3.25). Men in social class I marry women from that background seven times as often as if they chose irrespective of class; and so on. But this value is also affected by the size of the group. For example, class IIIM could not 'score' more than three on this measure because the group is so numerous that one quarter would marry within it even if marriage were random with respect to social class. But random mating by class is rather an unlikely event given the spatial clustering of social groups, social norms of eligibility, and the ways in which people meet their future partners—by physical proximity, at college, through recreation, at work—which all are already strongly selective by social class.

Data on marriage within and between classes can be used as well as the diagonal to construct measures of 'nearness' of each class to the others using

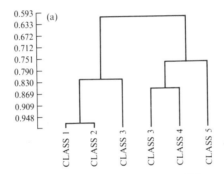

Hierarchical fusion of social classes by average linkage relatedness

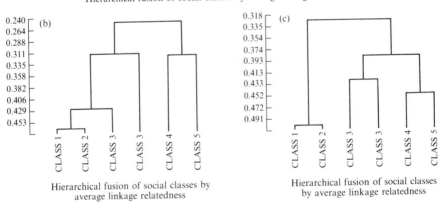

Hierarchical fusion of social classes by average linkage relatedness

Hierarchical fusion of social classes by average linkage relatedness

Fig. 3.3. Representation by cluster analysis (average linkage coefficient) of the pattern of similarity between social classes inherent in the relatedness matrix: (a) social mobility and marriage together; (b) social mobility only; (c) marriage only.

such techniques as cluster analysis; assuming of course that marriage is a good proxy for other forms of social interaction (Fig. 3.3). The relative isolation of class V and of classes I and II from the rest are salient features. The results from the analysis of this small sample are not consistent over different measures. But they suggest, as commonsense might indicate, that the perceived differences in status between adjacent classes is variable, depending among other things on the position of the observer himself. Table 3.26 measures relative isolation of social classes in a different way. It shows the time it would take for each social class to become indistinguishable from the others in terms of ancestry given the continuation of those patterns of marriage, starting from the (wholly arbitrary) assumption that they began completely distinct in terms of ancestry. On these data, the marriages between social classes are seen to be more effective than social mobility in homogenizing the origins of the population (assuming no selective effects in social mobility). Conclusions like this are typical of such enquiries in Western countries for which appropriate data are available. In such countries, patterns of marriage are probably the most open of any society where social stratification exists.

Most marriages follow from initial meetings in commonplace circumstances.

TABLE 3.26.

(a) Projection of effects of marriage alone upon social class relatedness. Number of generations required to take relatedness from 0 to 95%. First marriages.

		Social class				
		I	II	III (NM)	III (M)	IV
Social class	II	5				
	III (NM)	7	6			
	III (M)	9	8	7		
	IV	8	8	7	6	
	V	9	9	8	6	7

(b) Projection of effects of inter-generational social mobility alone upon social class relatedness. Number of generations required to take relatedness from 0 to 95%. First marriages.

		Social class				
		I	II	III (NM)	III (M)	IV
Social class	II	5				
	III (NM)	7	8			
	III (M)	10	10	8		
	IV	12	12	10	9	
	V	12	12	11	10	8

But these circumstances tend to be different for different social classes—manual workers are more likely to meet their future wives at dances and pubs, non-manual workers at work itself (the sex-ratio at the workplace of non-manual workers will usually be more favourable than that of manual workers), or at college. In particular, manual workers and especially men from class V are most likely to meet their future partners through simple proximity—in the street, public transport, or 'always known' (Coleman 1980). There is a well-known and longstanding relationship of distance between origins of marriage partners and social class, in town and in the country (Coleman and Haskey 1986; Harrison *et al.* 1971). Middle-class people are much more likely to marry a partner who lives at some distance and who was born far away.

CONCLUSIONS

It is possible to identify trends in social class differentials, especially in fertility, and also to identify a number of separate processes some of which are minimizing these differences, while others are tending to preseve them. High parity working-class fertility continues to decline, together with a reduction in unwanted births, while middle-class fertility holds steady. There is an increased concentration of family size on about two children in all classes, but the middle class are leading this trend. Fertility trends are still affected by the vestiges of the demographic transition, in so far as there is still room for the further development of education, particularly among children of manual workers, which might lead to less unwanted fertility and early marriage. All this means that social class differences are still contracting. Shorn of class I and V which are small in number, many of the tables in this paper do not show impressive demographic differences by social class. Away from the edges of society, social class differences across the greater part of the social spectrum are becoming less impressive, and increasingly related to differences in tempo. This may be regarded as the closing of a social gap in fertility which was opened wide in the 1870s and a return to the (probably) minor social differences in fertility before the 1850s; just as population growth has expanded, then contracted, as the demographic transition has progressed. In future, the major differences may be confined to class V—mostly unwanted high fertility—and classes I and II, with higher wanted fertility. Differences in sexual habits—in cohabitation, marriage, and divorce—can be expected to remain greater. This is because there can be positive advantages in advancing, as well as delaying, marriage in ways which there are not in fertility, also because modern marriage and its expectations are a severe test of social skills.

But other forces may ensure that social class differences do not disappear. The non-manual social classes delay their marriage and first birth ·much later than others, and have a much lower rate of illegitimacy. There are substantial reasons of personal advantage for doing so; rational responses to occupational differences, their earning curves and the time taken to acquire the qualifications to do the job. Given a general desire for at least a small number of children the existence of a positive gradient in fertility across classes I to III(NM) might be a permanent demographic feature. These groups share similar aspirations and role-models. If skilled manual workers come to share the same aspirations more, and have completed their transition to owner-occupation, then they may become part of the same continuum, at a position determined by their income level.

Some higher fertility in social classes V and IV may also be permanent because of the characteristics of some of the members of these social classes, the criteria on which individuals move socially and end up in particular occupations and social classes and because their behaviour is more likely to be affected by welfare and public housing policy. But any such differences may have decreasing demographic effect as non-manual occupations become more predominant numerically. The effect of high fertility of immigrant mothers may complicate this picture. Nothing is said here because there are few data on their position in the class structure (see *Employment Gazette* 1987). But their fertility and class position are both likely to change.

There are a number of reasons for supposing that social class differences may not disappear, or disappear fast. Some of them raise interesting questions about the structure of our society and the creation and transmission of its attitudes. There are cultural and psychological differences between social classes, as well as economic ones (Sissons 1970). Some of the 'ethnological' studies of social classes have described them in general terms (e.g. Klein 1965a; Mogey 1956; Hoggart 1957; Kerr 1958; Stacey 1960). These emphasize the concentration in working-class households of the 1950s and 1960s, especially those close to the level of subsistence and living on a day-to-day basis, of a number of attitudes or character traits. For example: of 'cognitive poverty', a habit of thinking rigidly and concretely, without speculation, over a narrow range of interests, which tend to maintain and perpetuate the traditional outlook characteristic of that section of society; a tendency to 'leave the field' or give up on difficult or long-term tasks or aims; a limited belief in ability to control their own lives. These contribute to distinctively different patterns of child rearing (Klein 1965b; see also Crandall and Crandall 1983) also noted in other cultures, although the patterns may not be constant over time (Bronfenbrenner 1966).

Demographic studies have discussed the importance of similar characteristics in the causation of early marriage and high and unplanned fertility (Askham 1975; Dunnell 1979), although without any formal psychological testing. Social psychologists have taken a considerable interest in social class differences in attitude and behaviour. Occupation, and occupational success itself appears to affect attitudes and personality (Andrisani and Nestel 1976). Adverse work experiences discourage belief in ability to control individual circumstances, or inclination to participate in the institutions of society. The rewards of employment, and not just work as such, encourage feelings of personal effectiveness and control among working women (Downey and Moen 1987). Social origin also affects behaviour and attributes which might affect social mobility, occupational selection, goals and their successful achievement (Hyman 1966; Kohn 1969).

One of the most important constructs is know as the 'locus of control' (Rotter 1966; Phares 1976), sometimes also known as 'mastery' or 'coping resources' (Pearlin and Schooler 1978), 'self-directedness' (Kohn and Schoenbach 1983), or 'personal efficacy' (Downey and Moen 1987). The location of control internally or externally to the individual is a measure of the extent to which that individual perceives success or failure as being contingent upon personal initiative. The more internal the location, the more worthwhile initiative becomes, in relation to (for example) income, housing, health, fertility, and marital success. The better educated and the better off are more likely to have effective coping responses, especially an essential variety of such responses (Pearlin and Schooler 1978; Chebat 1986). Manual workers are less inclined to believe they can influence their own or their families future and see less reason to plan activities far ahead or defer opportunities for gratification. These ideas may provide a means of analysing some social class differences in fertility, marriage, contraceptive use, and divorce, to see how far such differences should be regarded as rational responses to limited social and material opportunity, or instead as socially inherited handicaps, notably in adolescent fertility (Morrison 1985). So far, they have not been applied to questions of social class demographic differences. Neither have they been applied to questions of housing tenure or welfare. But the repeated claim that public housing encourages 'dependency' is an obvious candidate for testing in the terms of 'locus of control'.

There is also the separate question of ability. There are differences in average ability, however measured, by social class. Differences in ability affect the direction of social mobility and may in part drive the process. For example, within a given sibship, members with lower measured ability will tend to take lower-status occupations than those with higher ability (see Mascie-Taylor, this volume). Low ability is known to be associated with young childbearing and ineffective use of contraception (Kiernan and Diamond 1982). In the nature of the relatively undemanding occupations grouped into social class V, such characteristics may tend to be permanently overrepresented there by recruitment in each generation.

Policy on welfare and housing allocation may also tend to perpetuate some social demographic differences as long as it remains radically unchanged. Income support and housing priorities given to pregnant single females and to large families make it less rational to defer childbearing or marriage when earning power and the chances of owner-occupation may be low. Most people in this category have origins in social class V or IV. The scope for such influence is considerable. In 1988–89 £48 billion will be spent on social security transfer payments, about 20 per cent of GDP. Much of this, notably pensions, is unlikely to have demographic effects. But 450 000 families will receive Income Support (£8 billion in 1988–89), and 27 per cent of households are housed by local authorities or housing associations. The rationality of early marriage or childbearing outside marriage, where the normal housing expectation is provision by the local authority, has already been discussed. It is in the field of housing, rather than welfare, that radical change is most likely if the Government's recent (1987) policy to facilitate the transfer of public housing to alternative landlords achieves its objectives.

Non-material satisfactions of motherhood and the social recognition of its role, may make rational the sub-culture of illegitimacy of unqualified and almost unemployable girls who may have little capacity for any alternative outside dependency, given that welfare mechanisms exist to pay the cost of their fertility. Such an option depends upon welfare systems which give no preference to marital or to delayed fertility and may make lower-paid employment an economically unattractive option (Parker 1982).

Some of these points raise the question of whether social class demographic differences should be regarded as 'demographic regimes', and whether they can properly be described as part of a working-class 'culture', capable of being sustained and transmitted partly independently of occupation. In demography, 'demographic regimes' are a package of more or less rational responses adaptive to the particular social, economic, and housing expectations, and cultural values, of each group. This difficult question can only be raised, not solved, here. But it is important because it may determine the durability of demographic differences by social class, and the extent to which they might respond to changes in social policy. It also brings the question of social class differences into the broader intellectual context of the debate on the rationality of high fertility in other cultures. Early analysis regarded all Third World high fertility as necessarily irrational, at least on economic criteria. Later studies (Caldwell 1982) showed that this need not be so, although recent work has again emphasized the ideological rather than practical underpinning of high fertility.

For example, studies of traditional working-class life have emphasized the importance of supportive relations between a mother and her daughters, to compensate for the male and job-centered activities of her husband, to provide her with companionship in adult life rather than old age. This gives a rationale for a larger family size, produced early, which would be likely to include at least one daughter. Surveys have indeed shown some preference by mothers for daughters rather than sons for support in later life. Whether working-class households really gain material support in old age ('risk insurance') from their children is unlikely. Extended family households are not common in Western society in general, nor were they in the past (Smith 1986) whether for old age support or any other reason. And the housing conditions and income level of most working-class families do not make co-residence feasible.

This chapter has raised more question than it has answered. But at least it may have shown that social class differentials in fertility and marriage are a powerful indicator of substantial economic, social, and psychological inequalities in our society; that they are not likely to disappear fast; and that they will themselves contribute to the population dynamics of future generations.

REFERENCES

Ajami, I. (1976). Differential fertility in peasant communities: a study of six Iranian villages. *Population Studies*, **30** (3), 453–63.

Andorka, R. (1978). *Determinants of fertility in advanced societies*. Methuen, London.

Andrisani, P. J. and Nestel, G. (1976). Internal–external control as contributor to and outcome of work experience. *Journal of Applied Psychology*, **61** (2), 156–65.

Askham, J. (1975). *Fertility and deprivation: a study of differential fertility amongst working-class families in Aberdeen*. Cambridge University Press.

Banks, J. A. (1954). *Prosperity and parenthood*. Routledge, London.

Banks, J. A. (1981). *Victorian values: secularism and the size of families*. Routledge and Kegan Paul, London.

Barr, A. and York, P. (1982). *The official Sloane Ranger handbook*. Ebury Press, London.

Bastide, H., Girard, A., and Roussel, L. (1982). Une enquête d'opinion sur la conjoncture demographique. *Population*, **37** (45), 867–904.

Becker, G. S. (1981). *A treatise on the family*. Harvard University Press, Cambridge, Massachusetts.

Bendix, R. and Lipset, S. M. (ed.) (1967). *Class, status and power: social stratification in comparative perpective*, 2nd edn. Routledge and Kegan Paul, London.

Berent, J. (1954). Social mobility and marriage: a study of trends in England and Wales. In *Social mobility in Britain* (ed. D. V. Glass). Routledge, London.

Bernhardt, E. M. (1972). Fertility and economic status—some recent findings on differentials in Sweden. *Population Studies*, **26** (2), 175–84.

Blake, J. (1968). Are children consumer durables? *Population Studies*, **22**, 5–25.

Bone, M. (1978). *The family planning services: changes and effects*. HMSO, London.

Boston, G. F. P. (1984). *Occupation, industry, social class and socio-economic groups 1911–1981*. An unpublished working paper available from OPCS, Titchfield, Hants.

Bottomore, T. B. (1966). *Elites and society*. Pelican, Harmondsworth.

Britten, N. and Heath, A. (1984). Womens' jobs do make a difference. *Sociology* **18**, 475–90.

Britton, M. (1980). Recent trends in births. *Population Trends*, **20**, 4–8.

Bronfenbrenner, U. (1966). Socialisation and social class through time and space. In *Class, status and power*, (ed. R. Bendix and S. M. Lipset) pp. 362–76, Routledge and Kegan Paul, London.

Bulatao, R.A. and Lee, R.D. (ed.) (1983). Determinants of fertility in developing countries. Academic Press, New York.

Caldwell, J.C. (1982). *Theory of fertility decline*. Academic Press, London.

Calhoun, C.A and Espenshade, T.J. (1988). Childbearing and wives foregone earnings. *Population Studies*, **42** (1), 5–38.

Cartwright, A. (1978). *Recent trends in family building and contraception*. OPCS Studies in Medical and Population Subjects No. 34. HMSO, London.

Champion, AG., Green, AE., Owen, D.W., Ellin, D.J., and Coombes, M.G. (1987). *Changing places—Britain's demographic economic and social complexion*. Edward Arnold, London.

Chebat, J-G. (1986). Social responsibility: locus of control and social class. *Journal of Social Psychology*, **126** (4), 559–61.

Clarke, L., Farrell, C., and Beaumont, B. (1983). *Camden abortion study*. British Pregnancy Advisory Service, Solihull.

Cleland, J. and Wilson, C. (1987). Demand theories of the fertility transition: an iconoclastic view. *Population Studies*, **41**, 5–30.

Coleman, A. (1985). *Utopia on trial*. Hilary Shipman, London.

Coleman, DA. (1979). 'A study of migration and marriage in Reading, England in 1972-3'. *Journal of Biosocial Science*, **II**, 369–89.

Coleman, DA. (1980). The effect of socio-economic class, regional origin and other variables on marital mobility in Britain, 1920–1960. *Annals of Human Biology*, **8**, 1–24.

Coleman, DA. (1982). The population structure of an urban area in Britain. In *Current developments in anthropological genetics* (ed. M.H. Crawford and J.H. Mielke), Volume 2. Plenum Press, New York.

Coleman, D.A. (1984). Marital choice and geographical mobility. In *Migration and mobility: biosocial aspects of human movement* (ed. A.J. Boyce), pp. 19–56. Taylor and Francis, London.

Coleman D.A. and Haskey, J.C. (1986). Marital distance and its geographical orientation in England and Wales. *Transactions of the Institute of British Geographers*, **II**, 337.

De Cooman , E., Ermisch, J., and Joshi, H. (1987). The next birth and the labour market: a dynamic model of births in England and Wales. *Population Studies*, **41** (2), 237–68.

Crandall, V.C. and Crandall, B.W. (1983). Maternal and childhood behaviors as antecedents of internal–external control perceptions in young adulthood. In *Research with the locus of control construct*, Vol 2, Developments and social problems (ed. H.M. Lefcourt), pp. 53–103. Academic Press, New York.

Daunton, M.J. (1987). *A property-owning democracy?* Faber and Faber, London.

David, P.A. (1986). Comment on '*Altruism and the economic theory of fertility*' by G.S. Becker and R.J. Barro. In *Below replacement fertility in industrial countries* (ed. K. Davis, M.S. Bernshaw, and R. Ricardo-Campbell). Population and Development Review 12, Supplement. The Population Council, New York.

Davis, K., Bernshaw, M.S., and Ricardo-Campbell, R. (eds) (1986). *Below replacement fertility in industrial countries*. Population and Development Review 12, Supplement. The Population Council, New York.

Deane, P. and Cole, W.A. (1969). *British economic growth 1688–1959*, 2nd edn. Cambridge University Press.

Delvecchio Good, M-J., Farr, G.M., and Good, B.J. (1980). Social status and fertility: a study of a town and three villages in North Western Iran. *Population Studies*, **34** (2), 311–19.

Dennis, N., Henriques, F., and Slaughter, C. (1956). *Coal is our life*. Eyre and Spottiswood, London.

Department of the Environment (1969). *Council housing: purposes, procedures and priorities*. HMSO, London.

Department of the Environment (1977). *Housing policy: technical volume Part I*. HMSO, London.

Downey, G. and Moen, P. (1987). Personal efficacy, income and family transitions: a longitudinal study of women heading households. *Journal of Health and Social Behavior*, **28**, 320–33.

Dunnell, K. (1979). *Family formation 1976*. HMSO, London.

Employment Gazette (1987). Ethnic origin and economic status. *Employment Gazette*, January 1987, 18–29.

Ermisch, J. (1979). The relevance of the 'Easterlin hypothesis' and the 'New Home Economics' to fertility movement in Great Britain. *Population Studies*, **33** (1), 39–58.

Ermisch, J. (1983). *The political economy of demographic change*. Heinemann, London.

Ermisch, J. (1988). *Purchased child care, optimal family size and mother's employment: theory and econometric analysis*. Discussion Paper 238, Centre for Economic Policy Research.

Espenshade, T.H. (1984). *Investing in children: new estimates of parental expenditures*. Urban Institute Press, Washington, DC.

Furbank, P.N. (1985). *Unholy pleasure*. Oxford University Press.

Fox, J. and Goldblatt, P. (1982). Socio-demographic differences in mortality. *Population Trends*, **27**, 8–13.

Fussell, P. (1984). *Class: style and status in the USA*. Arrow, London.

General Household Survey 1984: General Household Survey 1986. London, HMSO (1988). London, HMSO (1986).

Glass, D.V. and Grebenik, E. (1954). *1946 Great Britain family census* Vol. 1. HMSO, London.

Glass, D.V. (1971). *Components of natural increase in England and Wales*. First Report from the Select Committee on Science and Technology, pp. 186–200. HMSO, London.

Goldstein, S. (1972). The influence of labour force participation and education on fertility in Thailand. *Population Studies*, **26** (3), 419–36.

Goldthorpe, J.H. and Hope, K. (1974). *The social grading of occupations—a new approach and scale*. Clarendon Press, Oxford.

Guttsman, W.L. (ed.) (1969). *The English ruling class*. Weidenfeld and Nicolson, London.

Hair, P.E.H. (1972). Children in society 1850–1980. In Barker, T. and M. Drake (ed.) *Population and Society in Britain 1850–1980*. Batsford, London.

Hall, J. and Jones, D.J. (1950). The social grading of occupations. *British Journal of Sociology*, **1**, 31–55.

Harrison, G.A., Hiorns, R.W., and Küchermann, CF. (1971). Social class and marriage patterns in nine Oxfordshire parishes. *Journal of Biosocial Science*, **3**, 1–12.

Haskey, J. (1983). Social class patterns of marriage. *Population Trends*, **34**, 12.

Haskey, J. (1984). Social class and socio-economic differentials in divorce in England and Wales. *Population Studies*, **38**, 419.

Haskey, J. (1986). Grounds for divorce in England and Wales—a social and demographic analysis. *Journal of Biosocial Science*, **18**, 127.

Haskey, J. (1987a). Divorce in the early years of marriage in England and Wales: results from a prospective study using linked records. *Journal of Biosocial Science*, **19**, 255–71.

Haskey, J. (1987b). Social class differentials in remarriage after divorce: results from a formal linkage study. *Population Trends*, **47**, 34–42.

Haskey, J. and Coleman, D.A. (1986). Cohabitation before marriage: a comparison of information from marriage registration and the General Household Survey. *Population Trends*, **43**, 15.

Heath, A. and McDonald, S-K. (1987). Social change and the future of the left. *The Political Quarterly*, **58** (4), 364–77.

Hinde, P.R.A. and Garrett, E.M. (1988). Work pattern, marriage and fertility in late 19th century England. *Journal of Family History*.

Hoggart, R. (1957). *The uses of literacy*. Chatto and Windus, London.

Hollingsworth, T.H. (1964). The demography of the British Peerage. *Population Studies*, **18**, Supplement.

Holmans, A.E. (1987). *Housing policy in Britain*. Croom Helm, London.

Holmans, A.E., Nandy, S. and Brown, A.C. (1987). Household formation and dissolution and housing tenure: a longitudinal perspective. *Social Trends*, **17** pp. 20–8. HMSO, London.

Honey, J. (1989). *Does accent matter?* Faber and Faber, London.

Hope, K. (1972). Marriage markets in the stratification system. In *The analysis of social mobility* (ed. K. Hope). Clarendon Press, Oxford.

Hudson, L. and Jacot, B. (1971). Marriage and fertility in academic life. *Nature*, **229**, 531–2.

Hull, T.H. and Hull, V.J. (1977). The relation of economic class and fertility: an analysis of some Indonesian date. *Population Studies*, **31** (1), 43–57.

Hyman, H.H. (1966). The value systems of different classes: a social psychological contribution to the analysis of stratification. In *Class, status and power* (ed. R. Bendix and S.M. Lipset), pp. 488–99. The Free Press, New York.

Ineichen, B. (1977). Youthful marriage—the vortex of disadvantage. In *Equalities and inequalities in family life* (ed. R. Chester). Academic Press, London.

Ineichen, B. (1979). The social geography of marriage. In *Love and attraction* (ed. M. Cook and C. Wilson), pp. 145–59. Pergamon Press, Oxford.

Innes, J.W. (1938). *Class fertility trends in England and Wales 1876–1934*. Princeton University Press.

James, W. (1974). Marital coital rates, spouses ages, family size and social class. *Journal of Sex Research*, **10** (3), 205–18.

Jones, P.R. (1982). Some sources of current immigration. In *The demography of immigrants and minority groups in the United Kingdom* (ed. D.A. Coleman), Academic Press, London.

Jones, E. and Grupp, F.W. (1987). *Modernisation, value change and fertility in the Soviet Union*. Cambridge University Press.

Joseph, Sir K. (1972). *The cycle of deprivation*. Speech to the Pre-School Playgroups Association, 29 June 1972.

Joshi, H. (1985). Motherhood and employment: change and continuity in post-war Britain. In *Measuring socio-demographic change* (British Society for Population Studies). OPCS Occasional Paper 34. OPCS, London.

Joshi, H. (1987). *The cash opportunity costs of childbearing: an approach to estimation using British data*. Centre for Economic Policy Research Working Paper No. 208. CEPR, London.

Kerr, C. (1958). *The people of Ship Street*. Routledge and Kegan Paul, London.

Kiernan, K.E. (1980). Teenage motherhood—associated facts and consequences—the experiences of a British birth cohort. *Journal of Biosocial Science*, **12**, 393–405.

Kiernan, K.E. and Diamond, I. (1982). Family of origin and educational influences on age at first birth: the experiences of the British birth cohort. *Centre for Population Studies Research Paper* no. 82–1. London School of Hygiene and Tropical Medicine, London.

Kiser, C.V. and Whelpton, P.K. (1950). Fertility planning and fertility rates by socio-economic status. In *Social and psychological factors affecting fertility* (ed. P.M. Whelpton and C.V. Kiser), Vol. 2. Milbank Memorial Fund, New York.

Klein, J. (1965a, b). *Samples from English cultures*, Vol. I, Vol. II. Routledge and Kegan Paul, London.

Kohn, M.L. (1969). *Class and conformity: a study in values*. Dorsey, Homewood, I.C.

Kohn, M.L. (1981). Personality, occupation and social stratification; a frame of reference. *Research in Social Stratification and Mobility: A Research Annual*, Vol. 1. JAI Press, Greenwich, CT.

Kohn, M.L. and Schoenbach, C. (1983). Class, stratification and psychological functioning. In *Work and personality: an inquiry into the impact of social stratification* (ed. M.L. Kohn and C. Schooler). Ablex Publishing Corporation, New Jersey.

Knodel, J. (1970). Two and a half centuries of demographic history in a Bavarian village. *Population Studies*, **24**, 353–76.

Kussmaul, A. (1981). *Servants and husbandry in early modern England*. Cambridge University Press.

Langford, C.M. (1976). *Birth control practice and marital fertility in Great Britain*. Population Investigation Committee, L.S.E, London.

Lee, R.D. and Bulatao (). *Income, Wealth and Demand for Children* (details to be confirmed)

Leete, R. and Fox, J. (1977). Registrar General's social classes: origin and uses. *Population Trends*, **8**, 1–7.

Lewis-Faning, E. (1949). *Report on an enquiry into family limitation and its influence on human fertility during the past fifty years*. Papers of the Royal Commission on Population, Volume 1. HMSO, London.

Macfarlane, A. (1986). *Marriage and love in England 1300–1840*. Basil Blackwell, Oxford.

Madge, N. (1988). Inheritance, chance and choice in the transmission of poverty. In *The political economy of health and welfare* ed. M. Keynes, D.A. Coleman, and N.H. Dimsdale. Oxford University Press.

Mant, D., Vessey, M., and N. Loudan (1988). Social class differences in sexual behaviour and cervical cancer. *Community Medicine*, **10**, 1, 52–6.

Marwick, A. (ed.) (1986). *Class in the 20th Century*. Harvester, London.

Martin, J. and Roberts, C. (1984). *Women and employment: a lifetime perspective*. HMSO, London.

Menefee, S.P. (1981). *Wives for Sale: an ethnographic study of British popular divorce*. Blackwell, Oxford.

Minford, P., Peel, M., and Ashton, P. (1987). *The housing morass*: regulation, immobility and unemployment. Institute of Economic Affairs, London.

Mitford, N. (ed.) (1956). *Noblesse oblige*. Penguin, Harmondsworth.

Mogey, J. (1956). *Family and neighbourhood: two studies in Oxford*. Oxford University Press.

Morris, R.N. and Mogey, J. (1965). *The sociology of housing*. Routledge and Kegan Paul, London.

Morrison, D.M. (1985). Adolescent contraceptive behavior: a review. *Psychological Bulletin*, **98** (3), 538–68.

Murie, A. (1974). *Household movement and housing choice*. University of Birmingham.

Murie, A. (1983). *Housing inequality and deprivation*. Heineman, London.

Murphy, M.J. (1985). Marital breakdown and socio-economic status: a re-appraisal of the evidence from recent British sources. *British Journal of Sociology*, **36**, 81–93.

Murphy, M.J. (1987). Differential family formation in Great Britain. *Journal of Biosocial Science*, **19**, 4635–8.

Murphy, M.J. and Sullivan, O. (1985). Housing tenure and family formation in contemporary Britain. *European Sociological Review*, **1**, 230.

Nag, M. (1968). *Factors affecting human fertility in non-industrial societies: a cross cultural study*. Yale University Publications in Anthropology No. 66, Human Relations Area Files Press. New Haven.

National Federation of Housing Associations (1985). *Inquiry into British housing; report, evidence*. NFHA, London.

Ni Bhrolchain, M. (1985). Birth intervals and women's economic activity. *Journal of Biosocial Science*, **17**, 31–47.

Ni Bhrolchain, M. (1987). Period parity progression ratios and birth intervals in England and Wales 1941–1971: a synthetic life table analysis. *Population Studies*, **41** (1), 103–26.

Notestein, F. (1963). Class differences in fertility. In *Class, Status and Power* (ed. R. Bendix and S.M. Lipset). The Free Press, Glencoe.

Office of Population Censuses and Surveys (1979). *Census 1971 England and Wales, fertility tables*, Volume II (10 per cent sample). HMSO, London.

Office of Population Censuses and Surveys (1980). *Classification of occupation 1980*. HMSO, London.

Office of Population Censuses and Surveys (1983). *Fertility report from the 1971 census*. Decennial Supplement DS No. 5. HMSO, London.

Office of Population Censuses and Surveys (1986). *Fertility trends in England and Wales 1975–1985*. OPCS Monitor FM1 86/2. HMSO, London.

Osborn A.F. and Morris, A.C. (1979). The rationale for a composite index of social class and its evaluation. *British Journal of Sociology*, **30**, 39–60.

Osborn, A.F., Butler, N.R., and Morris A.C. (1984). *The social life of Britain's five year olds: a report of the child health and education study*. Routledge and Kegan Paul, London.

Parker, H. (1982). *The moral hazard of social benefits: a study of the impact of social benefits and income tax on the incentives to work*. Research Monograph 37, Institute of Economic Affairs, London.

Pearlin, L.I. and Schooler, C. (1978). The structure of coping. *Journal of Health and Social Behaviour*, **19**, 12–21.

Phares, E.J. (1976). *Locus of control in personality*. General Learning Press, New Jersey.

Reid, I. (1981). *Social class differences in Britain*, 2nd edn. Grant McIntyre, London.

Rogers, J. (1984). *Poverty and population: approaches and evidence*. International Labour Organisation, Geneva.

Rosenfeld, E. (1951). Social stratification in 'classless' society. *American Sociological Review* **XVI** (6), 766–74. Reprinted in Tumin, M.M. (ed.) (1970): *Readings in social stratification*, pp. 113-22. Prentice-Hall, New Jersey.

Rotter, J.B. (1966). Generalised expectances for internal versus external control of reinforcement. *Psychological Monographs*, **80**, 1.

Royal Commission on Population (1950). *Papers of the Royal Commission on Population, Volume V. Memoranda presented to the Royal Commission*. HMSO, London.

Royal Society (1983). *Demographic trends and future university candidates—a working paper*. The Royal Society, London.

Safilios-Rothschild, C. (1982). A class and sex stratification theoretical model and its relevance for fertility trends in the developing world. In *Determinants of fertility trends: theories re-examined* (ed. C. Höhn and R. Mackensen). Ordina Editions, Liege.

Salt, J. (in press). Labour migration and housing in the UK: an overview. In *Labour markets and housing* (ed. C. Hamnett, and J. Allen). Hutchinson, London.

Schofield, M. (1968). *The sexual behaviour of young people*. Pelican, Harmondsworth.

Scott, J. (1982). *The upper classes: property and privilege in Britain*. Macmillan, London.

Searle, G.R. (1981). Eugenics and class. In *Biology, medicine and society 1840–1940* (ed. C. Webster). Cambridge University Press.

Simons, J. (1986). Culture, economy and reproduction in contemporary Europe. In *The state of population theory* (ed. D.A. Coleman and R.S. Schofield). Basil Blackwell, Oxford.

Smith, R.M. (1986). Transfer incomes, risk and security: the roles of the family and the collectivity in recent theories of fertility change. In *The state of population theory* (ed. D.A. Coleman and R.S.Schofield). Basil Blackwell, Oxford.

Stacey, M. (1960). *Tradition and change*. Oxford University Press.

Stevenson, T.H.C. (1920). The fertility of various social classes in England and Wales from the middle of the 19th century to 1911. *Journal of the Royal Statistical Society*, **LXXXIII**, 401–4.

Stevenson, T.H.C. (1923). The social distribution of mortality from various causes in England and Wales 1910-1912. *Biometrika*, **XV**, 382–400.

Stevenson, T.H.C. (1928). The vital statistics of wealth and poverty. *Journal of the Royal Statistical Society*, **XCI**, 207–30.

Stoeckel, J. and Chowdhury, A.K. (1980). Fertility and socio-economic status in rural Bangladesh: differentials and linkages. *Population Studies*, **34**, 519–24.

Stone, L. (1965). *The crisis of the aristocracy 1558-1641*. Clarendon Press, Oxford.

Stone, L. (1977). *The family, sex and marriage in England, 1500-1800*. Weidenfeld and Nicholson, London.

Stys, W. (1957). The influence of economic conditions on the fertility of peasant women. *Population Studies*, **II**, 136–48.

Susser, M.W. and Watson, W. (1971). *Sociology in medicine*, 2nd edn. Oxford University Press, London.

Szreter, S.R. (1984). The genesis of the Registrar General's social classification of occupations. *British Journal of Sociology*, **XXXV**, 522–46.

Thomson, G.H. (1946). *The trend of national intelligence*. The Galton Lecture 1946. Occasional Papers on Eugenics 3. The Eugenics Society, London.

Thornes, B. and Collard, J. (1979). *Who Divorces*? Routledge and Kegan Paul, London.

Wat, S-Y and Hodge, R.W. (1972). Social and economic factors in Hong Kong's fertility decline. *Population Studies*, **26** (3), 455–64.

Werner, B. (1984). Fertility and family background: some illustrations from the OPCS Longitudinal Study. *Population Trends*, **35**, 5–10.

Werner, B. (1985). Fertility trends in different social classes 1970 to 1983. *Population Trends*, **41**, 5–13.

Werner, B. (1988). Birth intervals: results from the OPCS Longitudinal Study 1971–1984. *Population Trends*, **51**, 25–9.

Westoff, C.F., Potter, R.G., and Sagi, P.C. (1963). *The third child: a study of prediction in fertility*. Princeton University Press.

White, P.E. (1985). Fertility and social class in a French village 1901–75. *Journal of Biosocial Science*, **17**, 253–65.

Willmott, P. and Young, M. (1973). *The symmetrical family: a study of work and leisure in the London region*. Routledge and Kegan Paul, London.

Wineberg, H. and McCarthy, J. (1986). Differential fertility in the United States, 1980: continuity or change? *Journal of Biosocial Science*, **18**, 311–24.

Woods, R.I. and Smith, C.W. (1983). The decline of marital fertility in the late 19th century: the case of England and Wales. *Population Studies*, **37**, 207–25.

Woolfe, M. (1971). *Family intentions*. HMSO, London.

Woolfe, M. and Pegden, S. (1976). *Families five years on*. HMSO, London.

Wrong, D.H. (1958). Trends in class fertility in Western Nations. *Canadian Journal of Economics and Political Science*, **XXIV**, 216–29. Reprinted in Tumin, M.M. (1970). *Readings on social stratification*, pp. 141–53. Prentice-Hall, NJ.

Wrong, D.H. (1967). *Population and society*. Random House, New York.

Young, M. and Willmott, P. (1957). *Family and kinship in East London*. Routledge and Kegan Paul, London.

Zweig, F. (1961). *The worker in an affluent society*. Heinemann, London.

4

The Biology of Social Class

C.G. NICHOLAS MASCIE-TAYLOR

INTRODUCTION

The primary objective of this chapter is to provide a comprehensive summary of those biological variables which are associated with social class. 'Associated' can have several meanings; it can refer to significantly more or less individuals than expected with a specified character in a social class. Alternatively, for a continuous character the term signifies average or mean differences between classes. In the latter case this might result in a measurable gradient for either an increase or decrease in mean values from social class I to V.

Social class has been defined here using the Registrar General's classification of occupations. This does not imply that this classification is superior to others. It reflects the situation whereby the majority of results come from official or semi-official research programmes which exclusively use this classification. Most of the associations discussed are drawn from British sources.

The chapter has a secondary aim; namely to try to account for the cause of the observed associations between some of the biological variables and social class. This is a much more difficult objective. For instance, suppose a decline in mean height from social class I to V was observed. Is this decline or gradient caused by genetic differences between classes, environmental differences, or some combination of the two? Is there some specific factor(s) within social class which is responsible for these observed differences? There is usually no easy answer to these questions but at least a realistic attempt will be made to further our understanding of the cause of these associations.

One case where the cause of the association has been reportedly found involves circumcision. This is an interesting example since the prevalence of circumcision was (a) found to show only a temporary difference between classes and (b) the difference could be due to an environmental factor. The information on circumcision comes from two cohort studies which commenced in 1946 and 1958. Each cohort examined all the children born in one week of the year and then studied them and their families from time to time. For instance the 1958 (National Child Development Study (NCDS) survey has undertaken follow ups in 1965, 1969, 1974, and 1983. These cohorts constitute random samples of the British population, the only bias being the time of year in which the study commenced.

By 11 years of age 22.7 per cent of the boys in the 1946 cohort had been circumcised and there was a significant class (simply divided into non-manual and manual groups) difference in prevalence. Whereas only 10 per cent of sons of heads of households working at manual tasks had been circumcised, 30 per cent of sons in the non-manual occupational class were circumcised. In the 1958 cohort, among boys of the same age only 10 per cent had been circumcised and there was little or no social class difference. Consequently two things had happened in the ten years between the cohorts: (a) a reduction in overall circumcision; and (b) an end to the social class difference.

Circumcision is generally performed during the first few weeks of life and there is usually no medical reason for such an operation (MacCarthy *et al.* 1952). The reduction in circumcision over this 10-year period was shown to be entirely due to the reduction in early circumcisions, i.e. those within the first year of life. The decline is proportionally greatest for children of non-manual workers, thereby accounting for the non-significant association with class in the 1958 cohort. Fogelman (1983) accounts for the elimination of the social class difference as well as the overall decline to paediatric opinion exerting itself through the hospitals, i.e. a simple environmental change. Although Fogelman might be correct, the chain of reasoning which says the effect of social class was environmental is a rather indirect argument, depending on disappearance of a difference and some plausible arguments about susceptibility to medical fashion.

SOCIAL CLASS DIFFERENCES IN MORBIDITY

Our understanding of the aetiology of many diseases including cancer, cardiovascular disease, and chronic locomotor disorders (Doll and Peto 1981) make it difficult to assess the precise role of social factors as causal agencies in spite of the overwhelming evidence of their importance. As Smith (1987) notes 'in most of the major diseases it is evident that social factors account for more of the variation than any factor after age.' As indicated a large number of conditions show social class differentials for both children and adults. In most cases there is a linear association with social class, the highest prevalence of various illnesses occurring in social class V. The range of these differences can be seen from Table 4.1 which presents the findings for the two extreme classes. Only adenoidectomy, hearing problems and eczema are found more commonly in the highest than in the lowest social class.

The results for adults also show major class differences. For instance the General Household Survey (GHS) obtained information by self-report on sickness (Hicks 1976). The relationship of self-reported health status to

TABLE 4.1.   Social class differences for various conditions or treatments.

Condition or treatment	Social class	
	I	V
Adenoidectomy	2.9%	1.0%
Low birthweight (<2500g)	4.4%	8.7%
Bed-wetting at 5 (1+ per wk)	6.8%	16.6%
Temper tantrums at 5 (1+ per wk)	5.9%	23.4%
Headaches (1+ per month)	4.3%	7.0%
Wheezing attack (ever)	17.1%	22.1%*
Eczema (ever)	16.6%	8.9%
Bronchitis (ever)	13.1%	20.8%
Pneumonia (ever)	1.1%	2.6%
Habitual mouth breathing or snoring (ever)	17.3%	20.3%
Suspected hearing problem (ever)	12.8%	6.6%

*Social classes IV and V combined.
Source: General Household Survey 1972.

clinically-defined disease is complex but the many studies that have compared self-reported illness with clinical examination have found without exception that disease is underreported (e.g. Cartwright 1959; Mechanic and Newton 1965). It is also important to remember that one is dealing with reports of disease and, in some respects, there may remain a question as to whether the same social class differences would be found if the conditions were evaluated by exactly comparable methods. It is also possible that some of the differences would disappear if the same independent measures were used for all classes.

Three different types of illness were considered; acute, chronic, and handicapping. GHS informants were asked the question 'During the two weeks ending last Sunday did you have to cut down at all on the things you usually do because of sickness and injury?' Those with chronic illnesses were also asked if it limited their activities compared with most people of their own age. The rates (per 1000 people) of self-reported illness by social class are summarized in Fig. 4.1. Acute illnesses show the least differentiation between classes and chronic the most—indeed the prevalence rate for handicapping chronic disease in social class V was 200 per 1000 people, well over twice the rate of 75 reported in class I. Rates for certain conditions are shown in Table 4.2. All show a lower rate in class I compared with V and in the majority of cases there are clines from class I to V.

One can also gain an idea of the well-being of different strata of the population from the extent to which they make use of health and health-related services. Again using the CHES data it is found that 23.4 per cent of five-year-olds from social class I are admitted to hospital compared to 29.3 per cent of children from class V. The hospital data corroborate the evidence from

C.G. Nicholas Mascie-Taylor

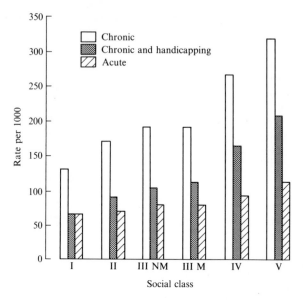

Fig. 4.1. Rates of self-reported illness by social class per 1000 people, men and women. *Source*: General Household Survey 1972.

TABLE 4.2. Rates per 1000 reporting certain conditions by social class.

Condition	Social class				
	Total	I and II	III (NM)	III (M)	I V and V
Mental disorders	11.0	6.5	7.1	10.4	19.3
Diseases of nervous system	8.7	6.5	7.6	6.9	13.6
Diseases of eye	7.3	7.1	8.2	5.6	7.7
Diseases of ear	7.7	4.3	5.4	8.5	10.9
Heart disease and hypertension	24.4	20.3	22.3	20.8	34.5
Other diseases of circulatory system	10.5	8.9	7.4	9.0	15.7
Bronchitis; acute, chronic, and unqualified	16.9	8.0	11.7	17.9	28.2
Other diseases of lower respiratory tract	13.5	10.5	12.9	12.5	19.0
Diseases of digestive system	11.4	8.1	8.9	11.5	16.9
Arthritis and rheumatism	27.5	20.1	24.7	21.8	44.2
Other diseases of musculo-skeletal system	9.3	7.1	9.4	9.2	11.3
Fractures, dislocation, and sprains	6.3	4.2	3.6	6.6	10.7
Other injuries, etc.	9.5	5.6	9.7	9.5	13.9

Source: General Household Survey 1972.

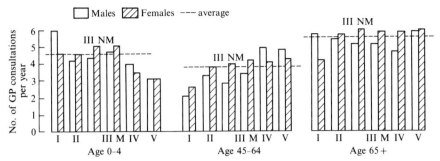

Fig. 4.2. Number of general practitioner consultations by social class per person per year.
Source: General Household Survey 1972.

Table 4.1 in suggesting greater illness among the lower social classes. The health behaviour also varied, with 7 per cent of social class V children never having been immunized to only 0.2 per cent for class I. For measles the figures were 52.6 per cent never immunized (class V) and 30.0 per cent (Class I) and for pertussis 13.5 per cent and 3.3 per cent (social classes V and I respectively). Just over 92 per cent of children of social class I mothers had been taken to a dentist by the age of five compared with only 55 per cent of children in social class V. Parents in the upper social classes might be more aware of the preventative health services than parents in the lower classes and hence make use of these services to a greater extent.

Figure 4.2 shows the average number of reported consultations per person per year with a general practitioner (General Household Survey 1972) by social class and for men and women separately at three age periods. For the over 65s there are hardly any social class differences in the use of primary medical care. The pattern in the 45–64 age group (deliberately chosen to exclude complications due to women's childbearing years) show a fairly regular rise in consultations as social class declines. For infants there is an obvious falling off in general practitioner consultations in the semi-skilled and unskilled groups.

The lower social classes need more medical help than the upper classes yet the evidence from both mortality (see Chapter 2) and morbidity is that they suffer more illnesses. What is the relationship between need and use—do the lower social classes use the primary health care services more than the upper classes? One way of measuring it is to compute the consultation index for each group (C I = consultation rate per 1000 people (in a two week period) divided by the amount of illness they reported). The results for chronic handicapping illness are shown in Table 4.3. As Blaxter (1976) noted, 'the higher social classes use them more in relation to the amount of illness they perceive themselves as suffering while the lower social classes may use their general practitioner more in terms of simple number'. A man in social class V who has a chronic illness visits his GP once compared to twice by the man from social class I; for women the differences are even more marked. The hospital consultation index (Table 4.3) shows a similar trend.

In summary it would appear that there are social class differences—some would say inequalities—in health and also in health care for both children and adults. Logan and Cushion (1974) have compared the standardized patient consulting ratio (actual number of patients consulting/expected number of patients consulting × 100) of a number of conditions for adult males by social class. The results (Table 4.4) highlight the differences between classes but also indicate that cardiovascular and psycho-neurotic conditions have above average morbidity in non-manual occupations while respiratory tract infections, gastric disorders, arthritis and rheumatism, and injuries are higher than average in manual occupations.

The social class differences for children based on general practitioner

TABLE 4.3. Consultation indices by social class.

Social class	General practitioner index		Hospital outpatient index	
	Male	Female	Male	Female
I	1.20	2.03	1.29	1.56
II	1.04	1.34	0.99	1.08
III(NM)	0.91	1.29	1.02	1.05
III(M)	0.84	1.20	0.93	0.84
IV	0.75	0.76	0.66	0.58
V	0.59	0.69	0.63	0.49

Source: Logan and Cushion 1974.

TABLE 4.4. Standardized patient consulting ratios by social class for selected diseases and conditions.

Social class	All diseases and conditions	Psycho-neurotic disorders	Acute naso-pharyngitis	Bronchitis	Arthritis and rheumatism
All social class	100	100	100	100	100
I	93	123	74	49	58
II	96	113	72	70	72
III	103	102	108	99	107
IV	99	82	111	116	115
V	99	90	104	146	111

Source: General Household Survey 1972.

TABLE 4.5. Conditions showing a social class relationship with respect to children.

Declining from social class I to V	Increasing from social class I to V	Highest in social class III
Whooping cough	Acute poliomyelitis	Acute nasopharyngitis
Asthma	Measles	Acute tonsilitis
Acute laryngitis	Infectious hepatitis	
Acute upper respiratory infection	Epilepsy	
Pyrexia of unknown origin	Bronchitis	
	Impetigo	
	Cough	
	Lacerations	
	Contusions	

Source: General Household Survey 1972.

returns are summarized in Table 4.5. There is high concordance between these results and the CHES national cohort study described earlier. Of the 16 conditions showing a social class gradient, five of them gave a gradient that declined in the direction of social class I to class V; nine of them gave a gradient that increased from class I to V, and two had the highest rate in social class III.

These social class differences are very marked. However one needs to be very careful in the interpretation of findings. As Golding (1986) wisely stated:

social class is a very odd variable—interpretation of social class differences varies from the assumption that one is detecting the results of 'deprivation' to the assumption that the social class trend is a clue to specific factors that are correlated with social class.

In other words social class does not necessarily define a unitary state but rather there is a 'social class complex'. Individuals in different social classes are likely to differ for a variety of characters which intercorrelate to varying degrees.

Golding attempted to detect the underlying components responsible for the association between class and pneumonia. For the CHES survey Golding reported ten features which showed significant class differences (Table 4.6). The largest differences were in teenage mothers, a six-fold difference; and mothers smoking 20+ cigarettes a day, a five-fold difference in frequency between classes I and V. Using these variables in a stepwise analysis, the pneumonia differences could be accounted for by the increased prevalence of parental smoking among the lower social classes combined with the fact that lower social class children are more likely to come from large families.

The fact that there remains no significant social class effect after other variables had been partialed out does not mean there is no social class association with pneumonia. What we see is primarily an association between smoking behaviour and pneumonia. But smoking behaviour associates with

TABLE 4.6.   Percentage of childen in social classes I and V reported to have various other background characteristics.

Characteristic	Social class	
	I	V
1. Teenage mothers (at time of birth)	2.2%	13.9%
2. Large families (4 + children when study child was 5)	8.7%	28.1%
3. One or both parents born outside the UK	7.8%	10.1%
4. Mother had paid job during child's life	47.0%	55.9%*
5. Resident in rural areas	23.5%	11.1%
6. Resident in poor inner-city areas	1.3%	20.9%
7. Mothers smoking when child aged <5	24.6%	56.2%
8. Mothers smoking 20+ per day	6.0%	29.7%
9. Fathers smoking when child aged 5	36.3%	67.9%
Fathers smoking 20+ per day	9.4%	37.6%
10. Child breast fed	59.2%	24.3%
11. Breast fed for 3+ months	25.6%	6.0%

*Social classes IV and V combined.
Source: Golding 1986.

social class—there is a linear increase in smoking between classes I and V. Consequently the results would suggest that the social class effect is operating through another variable, in this case smoking. Smoking behaviour is thus part of the 'social class complex' referred to earlier.

It remains unclear however exactly how smoking behaviour and pneumonia are related: it could be that smoking has a direct and adverse effect on the child—but equally the causal chain could be indirect, whereby for instance the social circumstances are more likely to lead lower social class mothers to smoke. These are only two of an extremely large array of potential answers. Consequently statistical 'explanations' although valid need not necessarily provide answers to the 'cause' of the observed pneumonia differences between classes.

A different approach was used by Kaplan and Mascie-Taylor (1985) when they examined biosocial factors in the epidemiology of childhood asthma. They used data collected through the National Child Development Study (1958 cohort).

Kaplan and Mascie-Taylor were concerned with predicting whether a child did or did not suffer from asthma. Consequently they examined a large number of background variables which had been shown to associate with asthma. Social class was one of these and in agreement with many other studies the NCDS data showed a preponderance of more asthmatics in non-manual homes. A stepwise discriminant function analysis statistically differentiated between asthmatics and non-asthmatics. Overall the discriminant analysis successfully classified 89.4 per cent of the sample and correctly predicted 52.3 per cent of those suffering from asthma. Social class was not one of the main predictors—these tended to be allergy-related variables and in particular eczema and hayfever. Again these results do not imply the absence of a social class effect—merely that there are other more powerful background characteristics.

MATING PATTERNS AND SOCIAL MOBILITY

Class endogamy and assortative mating

There are large bodies of data from both Britain and America which show that couples of similar backgrounds are more likely to marry than are couples from different backgrounds. In particular couples of similar social class are more likely to wed than couples who are of more dissimilar class. The data from the NCDS survey have been chosen to illustrate this finding (Table 4.7). In this case the social class similarity was based on wives who were working at the time of their marriage. Even higher levels of similarity occur when the wife's father's occupation is used.

We therefore have strong evidence for class endogamy. Meanwhile geneticists and biological anthropologists have found evidence for positive assortative marriage and mating for continuous characters like height and IQ. Positive assortative mating refers to the situation in which like preferentially mates with like for some phenotypic character(s). If the characters are hereditary, assortative mating can have profound effects on the genetic structure of populations; heterozygosity at those loci responsible for the character will decrease with a concomitant increase in homozygosity.

IQ and height both associate with social class; there is a tendency for mean heights (and IQs) to decline from social class I to V. Consequently when class endogamy occurs there is an increased likelihood of spouses being more

TABLE 4.7. The frequency of spouses by social class.

Husband's social class	Wife's social class					
	I and II	III (NM)	III (M)	IV	V	Total
I and II	192	311	64	58	10	635
III(NM)	68	230	46	60	13	417
III(M)	145	645	362	427	112	1291
IV	6	20	5	22	9	62
V	41	212	178	274	82	787
Total	452	1418	655	841	226	3592

Source: Mascie-Taylor 1987.

similar in height and IQ. Class endogamy could therefore account for much of the observed positive assortative mating for IQ and height. Consequently it is unclear to what extent the observed levels of assortative mating are due to active choice of partners or to what extent they are due to class endogamy.

Mascie-Taylor and Vandenberg (1988) attempted to clarify this situation by differentiating the broader effects of propinquity from specific mate selection based on personal preference *per se*. Using data from a Cambridge family study (Mascie-Taylor 1977) they studied the similarity between 193 husband and wife pairs for three personality variables (extroversion–introversion, neuroticism, and inconsistency) and two IQ components (verbal and non-verbal). Extroversion–introversion and neuroticism are two personality factors often measured by psychological tests (in this case Eysenck's Personality Inventory was used). IQ is often subdivided into various components depending on the test used. The Wechsler IQ test divides total IQ into these two subcomponents.

The majority of these tests show social class differences. For instance, manual workers are more likely to be extrovert and inconsistent than are members of non-manual occupations. Both IQ components associated with class, verbal more so than non-verbal IQ.

The passive elements, defined as those resulting from educational, social, and geographical propinquity accounted for two-thirds of the observed spousal association for IQ. However the remaining third which was ascribed to personal choice was still significant for verbal IQ but not for non-verbal IQ. For the personality variables the initial correlations were low but even so the personal choice component was still significant.

More elaborate attempts to untangle active and passive elements in mate choice have been carried out reanalysing the Cambridge and Otmoor data sets (Mascie-Taylor *et al.* 1987; Mascie-Taylor 1987; Mascie-Taylor 1988). Using stepwise multiple regression, the effects of years of education, social class, type of school attended, geographical propinquity (simply defined as locally or non-locally born), birth order, and family size were independently removed from husband and wife IQ scores after which the residual spousal correlation was computed. In theory these analyses remove the main background factors so that any residual association between spouses should reflect personal choice. Only the residual verbal IQ component remained significant (which is consistent with the results obtained by Mascie-Taylor and Vandenberg) albeit with a greatly reduced value—down from an initial coefficient of $+0.465$ to a residual correlation of $+0.107$.

Similar analyses have been carried out for height and weight. Reported statures and weights were available for mothers and fathers of children in the NCDS (1958) cohort. The overall correlation for stature was $+0.277$ and for weight $+0.115$. Both these coefficients are similar to values obtained from measured stature and weight (Roberts 1977). Using the stepwise regression method the effects of social class, educational background, region, and age were removed independently and the residual stature and weight correlation coefficients computed. Both remained highly significant,—indeed the effects of the background variables on spousal likeness were small—the stature correlation fell by 0.015 and weight not at all.

At face value these results suggest that much of the perceived assortative mating for IQ and personality might be 'explained' by homogamy for other variables including class endogamy. However for stature and weight the assortative mating levels appear high even after 'correction' for social class, etc. It must be remembered that the height and weight data examined here were reported at least 11 years after marriage. One would not expect height to show enormous variation over time but weight usually increases during marriage and it is unknown to what extent the correlation reflects that at marriage. Furthermore only a limited number of background variables were considered and there might well be some which would reduce the residual spousal correlation.

Social mobility

We have a picture of assortative mating occurring for height, weight, and IQ partly because of class endogamy and partly through additional personal choice. What maintains the height, weight, and IQ clinal association with social class since it is known that class endogamy although high is not absolute (see Table 4.7)? The answer appears to be selective migration through social mobility.

In Britain there is a large amount of movement between social classes. Burt (1961) initially suggested that differences between the general intelligence of children and the average values for the social class into which they were born

TABLE 4.8. Social mobility of sons and mean difference in IQ between them and their fathers.

Number of steps	Cambridge				Otmoor			
	N	Verbal	Visuo-spatial	Total	N	Verbal	Visuo-spatial	Total
Upwardly mobile								
+1	15	+4.3	+4.5	+3.9	14	+2.9	+3.5	+3.2
+2	14	+7.1	+10.0	+8.7	9	+6.3	+7.2	+6.2
+3	6	+11.7	+14.2	+13.7	—	—	—	—
Downwardly mobile								
−1	10	+0.40	−9.8	−4.8	20	−6.2	−2.5	−3.5
−2	6	−8.7	−15.8	−13.3	4	−5.8	−12.8	−9.3
−3	1	−10.0	−22.0	−15.0	1	−17.1	−21.9	−20.9

Source: Mascie-Taylor 1977 and Gibson *et al.* 1983.

would produce a large and fairly constant amount of basic mobility. Work by Gibson (1970), Gibson and Young (1965), and Waller (1971) substantiated Burt's hypothesis. For instance, a son born into social class III with a high IQ was more likely to be upwardly socially mobile; conversely a son from a high social class with a lower IQ was more likely to migrate down the social classes.

Later work by Mascie-Taylor and Gibson (1978) and Gibson *et al.* (1983) looked at the extent of mobility and the differences in verbal IQ and non-verbal IQ between fathers and their sons. Using data from the Cambridge (Mascie-Taylor 1977) and Otmoor studies (Harrison *et al.* 1974) they showed that there were significant regressions of extent of mobility and IQ difference between fathers and sons (Table 4.8). Although the numbers are not large ($n = 52$ Cambridge and $n = 85$ Otmoor) there does seem a relationship between the magnitude and direction of the father–son difference and the extent of mobility. If these results were extrapolated to the population then they would be expected either to promote or maintain social class differences. There is of course some question as to whether intellectual abilities help determine social class as well as social class helping to determine IQ.

GENETIC DIFFERENCES BETWEEN SOCIAL CLASSES

Blood groups

The observed IQ–class differences appear to be partly maintained by assortative mating and selective migration. IQ (and height and weight) has a significant heritability although the magnitude of the heritability is still disputed. Consequently there is the possibility that the IQ differences between classes might be caused by genetic differences between them. What is the evidence for such differences?

In 1973, Gibson *et al.* reported an IQ–ABO blood group association using data from Otmoor (Harrison *et al.* 1974). On average, individuals of blood group O scored higher on IQ tests than individuals of blood group A. The results were confounded by the heterogeneous nature of the population in Otmoor comprising as it did a mixture of locally and non-locally born. The non-locally born mean IQ was considerably elevated above the locally born and there were also gene frequency differences between local and non-locally born. However if their IQ–ABO findings were generalized to the British population one might expect to find more blood group O people in the higher social classes and more individuals with blood group A in the lower classes (Table 4.9).

Recently Beardmore and Karimi-Booshehri (1983) examined the distribution of ABO blood groups in a large British blood donor sample (Kopec 1970) and found an association between ABO blood group phenotype and social

TABLE 4.9. Mean I Q's and A_1 and O phenotypes among locally and non-locally born males and females living in Otmoor.

	Female		Male	
	A_1	O	A_1	O
Local	99.3	100.0	100.5	103.4
Non-local	107.5	110.2	113.3	113.0

Source: Harrison *et al*. 1974.

class. In both sexes the A phenotype was significantly more frequent (and the O-phenotype less frequent) in social classes I and II and the converse for classes III, IV, and V. These results are in the opposite direction to those predicted from the Otmoor study. The report by Beardmore and Karimi-Booshehri published in *Nature* generated 'a wealth of correspondence'—actually five published replies. Two of the papers provided detailed analyses of the 1958 and 1970 cohort studies. Neither Mascie-Taylor and McManus (1984, analysing 1958 cohort) nor Golding *et al*. (1984, analysing 1970 cohort) detected any association between ABO blood group distribution and class or between socially mobile groups and class.

It seems premature, given the existing evidence, to suggest that ABO blood group differences exist between classes. The editors of *Nature* were also undecided since they wrote, when discussing the ABO/class association 'The general conclusion seems to be that there can be no general conclusion'. Of course the suggestion that genetic differences exist between classes is far from novel (e.g. Herrnstein 1971). There may be genetic differences between classes, other than in the frequencies of certain blood groups, but empirical evidence is still awaited.

Schizophrenia, alcoholism, and biological determinism

The extent to which psychoses and alcoholism are inherited has generated enormous discussion. Twin studies have without exception yielded positive (genetic) results as have adoption studies. This is not the forum for a discussion on the nature/nurture of schizophrenia, alcoholism, and other disorders. What is relevant is that schizophrenia and to some extent alcoholism are most often diagnosed in the working class, inner-city dwellers and least often in middle and upper class suburban dwellers.

To a social theorist the argument about the social context that determines the diagnosis is clear. An example of the class nature of the diagnosis of mental illness comes from the studies of depression in Camberwell. Brown and Harris (1978) showed that about a quarter of working-class women with children living in Camberwell were suffering from what they defined as neurosis, mainly severe depression, compared to only 6 per cent in the middle-class women. A large proportion of these women—who would have been diagnosed as ill and hospitalized had they attended a psychiatric clinic—had suffered 'severe threatening events' (loss of husband or economic insecurity) within the year prior to the study. The use of drugs, mainly tranquilizers, was also high. However biological determinists would counter the social evidence by arguing that people with genotypes predisposing toward schizophrenia may drift downward in occupation and living conditions until they find a niche most suited to their genotype. As Rose *et al.* (1984) state: 'it would be a brave biological determinist who would want to argue that in the case of the depressed housewives of Camberwell it was their genes at fault'.

Obesity

There is a large amount of evidence which points to a greater prevalence of obesity in the lower social classes in highly industrialized societies. Although children belonging to the lower social classes might well, on average, be lighter than upper social class children they are much heavier when their height is taken into account. Such studies make use of the National Center for Health Statistics (NCHS, USA) or Harvard standards to obtain weight-for-height of a child compared with the norms from a cross-section of the United States' population. Although weight-for-heights are only suggestive, more conclusive evidence has been obtained from studies which also measured skinfold thickness. These studies, conducted for example in London and New York found obesity to be more common in the poor than in the well-off, upper social groups. There is evidence from a study in Kent, England that although children of lower social classes had a lower total nutrient intake than higher-

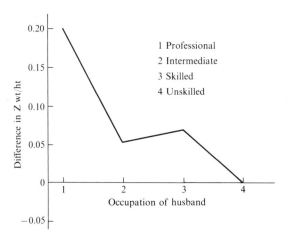

Fig. 4.3. Differences in nutritional status by occupation of husband (unskilled occupations set to zero).
Source: Nestel and Mascie-Taylor 1987.

class children, they consumed carbohydrate and added sugar equal to those of the higher social class so that the quality of the lower-class diet was worse (Cook *et al.* 1973). It would seem that the higher prevalence of obesity in the lower classes begins early and is related to diet.

The higher prevalence of obesity in the lower social classes applies only to countries where calorie intake is adequate for all social/income groups. In developing countries the opposite results are obtained. For example recent work in Sudan shows the prevalence of obesity as measured by weight-for-height appears only in the affluent upper social groups (Nestel and Mascie-Taylor 1987). It can be seen from Fig. 4.3 using weight-for-height that the higher values are found predominantly in the professional (classes I and II) groups. There are numerous other examples where similar findings have been obtained.

Age at menarche

Progress to sexual maturity is assessed by stages of appearance of secondary sexual characteristics. By far the most common indicator in girls has been age of menarche. Consideration of the relationship between age at menarche and social class must take into account the method of collection of the data. Unfortunately the easiest method, restrospective assessment, is the most unreliable; the other methods are status quo (most reliable) and prospective (next most reliable). Since it is easier to do a retrospective study it is not surprising that this method has generated a considerable literature.

Recent research suggests in the main that there is little or no association between social class and age at menarche. For instance using the NCDS database Mascie-Taylor and Boldsen (1986) found that there was a just significant association between class and menarcheal age (retrospective) with a tendency for it to increase from social class I to V; whereas there was marked regional variation with lower ages in Central and Southern England and higher ages in Northern England, Wales, and Scotland. Since class and region are correlated they also examined the interaction between class and region which was insignificant. Roberts *et al.* (1986) failed to show a social class effect in Cumbrian data, a result which parallels that of work in South Shields, Northumberland, Swansea, and Warwick. Only the results from Newcastle-upon-Tyne which divided the data into non-manual and manual categories continued to show a social class effect.

However Bielicki *et al.* (1986) published results for 19 000 Polish schoolgirls from three large cities of the Upper Silesia conurbation. They studied the relationship between menarcheal age in relation to parental education and father's occupation. Age at menarche tended to increase with decreasing parental education although the gradient was not steep. Mean menarcheal age ranged between 12.82 and 13.30 years for different occupational groups (Fig. 4.4).

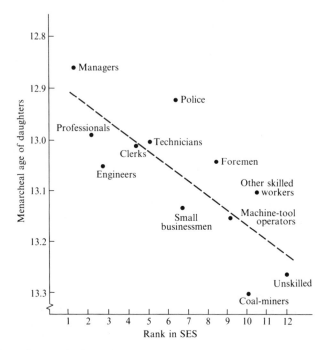

Fig. 4.4. Age of menarche by socio-economic status.
Source: Bielicki *et al.* 1986.

It is unclear therefore whether the British or Polish data are the more typical. It could be that the Polish results reflect the social inequality prevailing in the population a decade before their study was conducted. The increasing rate of obesity in manual-class families as they approach maturity suggests that in developed societies these females are as much or more in positive calorific balance than women of the non-manual classes among whom, indeed, dieting and even overdieting may be common (anorexia is much more common in middle-class girls). So it is hardly surprising that the age at menarche, which may have been differentiated by social status (as it is in some societies), has converged to similar low ages among social classes of some communities in economically developed countries.

Skeletal maturation

There is some evidence that skeletal maturation is slower in lower socio-economic compared with middle and upper groups. An example is presented in Fig. 4.5. The data are from Hong Kong and skeletal maturation was assessed using the Greulich-Pyle standards (Low *et al.* 1964). Also apparent, though not shown here, was the characteristic Asiatic increase in rate at adolescence

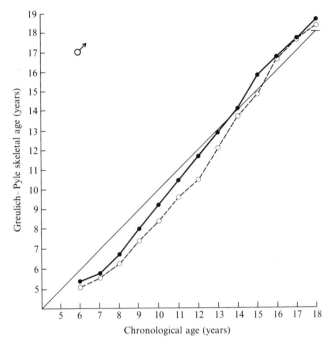

Fig. 4.5. Mean skeletal age of boys in high (●——●) and low (○--○) social groups in Hong Kong.
Source: Low *et al*. 1964.

which began a year earlier in the upper-class boys and earlier still in the upper-class girls. These differences can be accounted for by environmental rather than genetic differences.

IQ

The existence of a relationship between IQ score and social class has already been mentioned (see p. 125). Numerous studies in both this country and in other parts of the world have shown substantial correlations between parental social class and both the IQ and scholastic attainment of their children (Davie *et al*. 1972). It is worth pointing out that the mean difference between social class I and V in IQ is of the order of 15 IQ points—which is about the same as the black/white US IQ difference. Consequently class differences appear to be accounting for a large and significant proportion of the observed variance of IQ.

Furthermore several studies have shown that mild retardation mainly occurs in families in which the father has an unskilled or semi-skilled manual occupation. Mild retardation is in addition particularly common when there is poverty, family disorganization, overcrowding, and a large number of children in the family. In these circumstances it is quite common for several of the children to be mildly retarded or to have an IQ score near the bottom end of the normal range (Rutter and Madge 1976).

The direct way in which social class influences could be demonstrated is to use adoption studies by determining the association between parental social class and a child's IQ within a sample of adopted children in which selective placement has not occurred. Data which approach this ideal are available from some early American studies, for instance those of Leahy (1935), Burks (1928), and Skodak and Skeels (1949). These studies indicate a low but positive association between adoptive father's occupational status and child's IQ.

However we know that a large number of other variables also associate with IQ—for instance, extent of crowding—which is itself not independent of social class. It is unclear therefore to what extent the difference in IQ is due to class *per se* or to what extent it is confounded by other variables.

This point was forcibly made by Jones and Cameron (1986) who attacked the simplistic use of social class analysis. They were particularly concerned with health and health-related conditions, but their view that social class is an embarrassment to epidemiology is equally valid in other contexts and for other analyses. They said that 'It [social class] obscures much that would be of value and should be abandoned for scientific enquires'. They conclude by saying:

If what is required is an analysis of society showing the importance of some circumstance which society can change for the better, and about which we have a theory on the genesis of disease, then we should make an analysis of that circumstance. We should identify in each case the status of each individual's educational achievement, nutritional level, degree of overcrowding in housing, or whatever condition in which we are interested. If the analysis supports our theory, we should make the appropriate recommendations for change.

McManus and Mascie-Taylor (1983) and Mascie-Taylor (1983) have examined the association between a large number of social, economic, family variables, and IQ score of children aged eleven in the NCDS cohort. In the more extensive analysis Mascie-Taylor examined 29 variables in relation to IQ. The results are presented in Fig. 4.6. The upper line shows the relationship between mean IQ and social class *per se* and the usual downward trend from class I to class V is apparent. The mean difference between class I and V exceeds 17 points. After multiple regression analysis, which removed the effects of the 28 variables and higher order terms, a substantial IQ/class association still remained and the mean difference between social classes I and V still exceeded ten IQ points.

It would thus appear from these indirect analyses that even after removal of the more obvious features which fall into the 'social class complex', a substantive residual association remains. Such differences might reflect genetic differences between social classes. Equally the differences could reflect

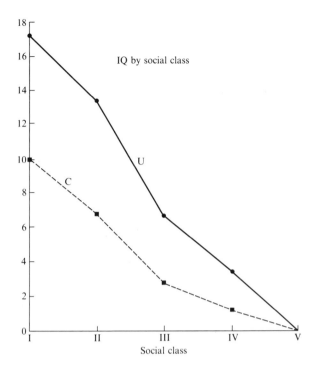

Fig. 4.6. Relationship between IQ of children and parental class; the IQ score of social class V has been set to zero. U = uncorrected scores; C = scores after removal of controlling variables.
Source: Mascie-Taylor 1983.

untested environmental variables and imperfections in the measurement of variables that were analysed. What is important to stress is that these results are not in accord with Jones and Cameron's hypothesis since they would expect to find the near elimination of a social class effect. It may still be true, of course, that the moiety which is susceptible to social action is more 'important' than the residual of social class which is of genetic or unknown cause.

Height

There is considerable evidence for social class heterogeneity with respect to height. For example Boldsen and Mascie-Taylor (1985) examined the heights of the 33 000 individuals comprising fathers, mothers, and children who took part in the NCDS survey. Overall results for all three family positions (Fig. 4.7) clearly show a linear trend with decreasing height from class I to class V. Indeed the mean difference in height between class I to V was close to 2

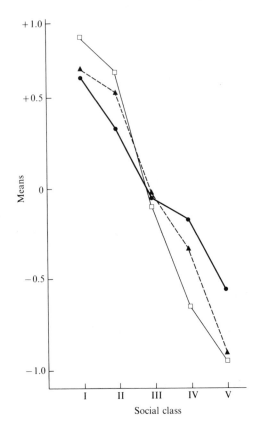

Fig. 4.7. Mean heights for each family position by social class as deviations (inches) from the grand mean. (□ = fathers, ● = mothers, and ▲ = children).
Source: Boldsen and Mascie-Taylor 1985.

inches. They also examined the regional variation and found clear differences (Fig. 4.8) with Scotland having the lowest overall mean and the Southern regions comprising London, East Anglia, the South and South East the highest. However, within each of the eleven standard areas the social class gradient of a decline in mean height from class I and V was replicated. Analysis of variance showed the social class differences were much larger than regional heterogeneity. The maximum difference between regions was only 0.91 inches compared to nearly 2 inches for social class. Cluster analysis of the eleven regions and five social classes, i.e. 55 cells also indicated that the major differentiation was between an upper social class segment (cluster 1, Fig. 4.9) comprising all eleven regions of social class I and most of social class II and a remaining middle- and lower-class segment. Scottish residents in social class V

Fig. 4.8. Mean heights of children by region as deviations from the grand mean.
Source: Boldsen and Mascie-Taylor 1985.

were shortest on average and they did not cluster with either of the two main segments.

More detailed analysis using the NCDS cohort have recently been carried out (Terrell and Mascie-Taylor 1990) similar to those reported for IQ. After the removal of 11 other variables, a significant social class effect still remained (Fig. 4.10) although the differences in means between classes I and V fell by approximately two-thirds. Again it is unclear to what extent these differences are the result of other environmental factors, genetic differences, or a combination of the two.

Furthermore, Lasker and Mascie-Taylor (1988) have shown that these social class differences are established early (largely before the age of seven); there was no significant influence of social class on growth in stature between ages 11 and 16. They also found that families who improved their social standing

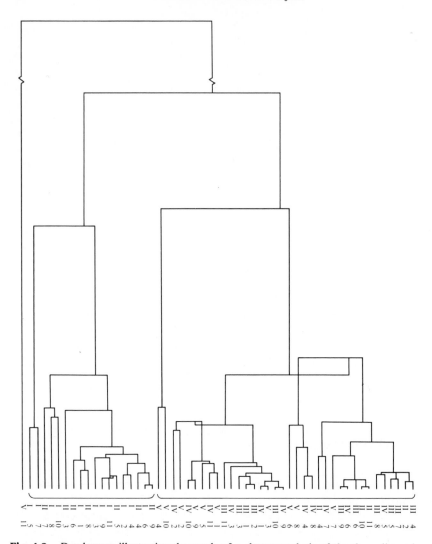

Fig. 4.9. Dendogram illustrating the result of a cluster analysis of the three-dimensional means in each of the 5.5 cells: Social class (I to V); regions (1 to 11).
Source: Boldsen and Mascie-Taylor 1985.

TABLE 4.10.   Head circumference by vocational status.

Vocational status	N	Mean (in mm)	S.D.
Professional	25	569.9	1.9
Semi-professional	61	566.5	1.5
Clerical	107	566.2	1.1
Trades	194	565.7	0.8
Public service	25	564.1	2.5
Skilled trades	351	562.9	0.6
Personal services	262	562.7	0.7
Labourers	647	560.7	0.3

Source: Gould 1981.

had children who had already previously exceeded the average height for their previous social class and the reverse was observed among the downwardly mobile.

Head circumference

Hooton (1939) studied the head circumference of white Bostonians as part of his study of criminals. Epstein (1978) reworked Hooton's data and purported to show 'the ordering of people according to head size yields an entirely plausible ordering according to vocational status'. The results are presented in Table 4.10. However as noted by Gould (1981) there are serious errors and omissions in Epstein's Table. Three trades—factory workers, transportation employees, and extractive trades (farming and mining) were excluded and all three had mean head circumferences above the grand mean for all the professional groups! Hooton also thought there was little or no relationship between head size and professions.

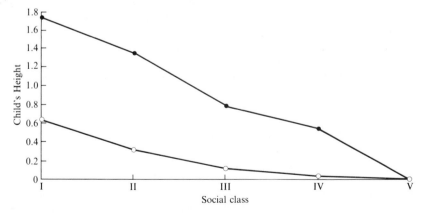

Fig. 4.10.   Uncontrolled (●——●) and controlled (○——○) child's height by social class.
Source: Terrell and Mascie-Taylor 1990.

CONCLUSIONS

Thus there are differences in many, but not all, human biological variables among social classes. Some variables, such as early childhood growth, are associated with father's type of occupation in all sorts of circumstances. Others, such as adult weight, vary among populations—lower social classes being relatively heavy in developed countries, but light where there is nutritional stress. Some of the variation is explained by specific cultural and other environmental factors that are associated with social class.

There is, however, a fraction of the social class variation in psychological and physical traits that remains after allowing for such factors. Efforts to partition this remaining fraction among other environmental conditions, errors in measurement, and any underlying genetic differences remain so far largely unresolved.

REFERENCES

Beardmore, J.A. and Karimi-Booshehri, F. (1983). ABO blood groups and social class. *Nature*, **303**, 522–4.

Bielicki, T., Waliszko, A., Hulanicka, B., and Kotlarz, K. (1986). Social-class gradients in menarcheal age in Poland. *Annals of Human Biology*, **13**, 1–11.

Blaxter, M. (1976). Social class and health inequalities. In *Equalities and inequalities in health* (ed. C.O. Carter and J. Peel). Academic Press, London.

Boldsen, J.L. and Mascie-Taylor, C.G.N. (1985). Analysis of height variation in a contemporary British sample. *Human Biology*, **57**, 473–80.

Brown, G.W. and Harris, T. (1978). *Social origins of depression; study of psychiatric disorder in women*. Tavistock, London.

Burks, B.S. (1928). The relative influence of nature and nurture upon mental development; a comparative study of foster parent–foster child resemblance and true parent–true child resemblance. *Yearbook of National, Social and Education*, **27**, 219–316.

Burt, C. (1961). Intelligence and social mobility. *British Journal of Educational Psychology*, **14**, 1–28.

Cartwright, A. (1959). Problems in the collection and analysis of morbidity data. *Milbank Memorial Fund Quarterly*, **37**, 33–46.

Cook, J., Altman, D.G., Moore, D.M.C., Topp, S.G., Holland, W.W., and Elliott, A. (1973). A survey of the nutritional status of schoolchildren. Relation between nutrient intake and socio-economic factors. *British Journal of Preventive and Social Medicine*, **27**, 91–9.

Davie, R., Butler, N.R., and Goldstein, H. (1972). From birth to seven: the second report of the National Child Development Study (1958 cohort). Longman and the National Children's Bureau, London.

Doll, R. and Peto, R. (1981) *The causes of cancer*. Oxford University Press.

Epstein, H.T. (1978). Growth spurts during brain development: implications for educational policy and practice. In *Education and the brain* (ed. J.S. Chall and A.F. Mirsky) University of Chicago Press.

Fogelman, K. (1983). *Growing up in Great Britain*. Macmillan, London.

General Household Survey (1972). HMSO, London.

Gibson, J.B. (1970). Biological aspects of a high socio-economic group. 1. IQ, education and social mobility. *Journal of Biosocial Science*, **2**, 1–16.

Gibson, J.B. and Young, M. (1965). Social mobility and fertility. In *Biological aspects of social problems* (ed. J.E. Meade and A.S. Parkes) Oliver and Boyd, Edinburgh.

Gibson, J.B., Harrison, G.A., and Hiorns, R.W. (1973). IQ and ABO blood groups. *Nature*, **163**, 873–9.

Gibson, J.B., Harrison, G.A., Hiorns, R.W., and Macbeth, H.M. (1983). Social mobility and psychometric variation in a group of Oxfordshire villages. *Journal of Biosocial Science*, **15**, 193–205.

Golding, J. (1986). Cross-cultural correlates of ill-health in childhood. In *The health and development of children* (ed. H.B. Miles and E. Still) The Eugenics Society, London.

Golding, J., Hicks, P., and Butler, N.R. (1984). Blood group and socio-economic class. *Nature*, **309**, 396.

Gould, S.J. (1981). *The mismeasure of man*. Penguin Books, London.

Harrison, G.A., Gibson, J.B., Hiorns, R.W., Wigley, J.M., Hancock, C., Freeman, C.A., Kuchemann, C.F., Macbeth, H.M. Saatcioglu, A., and Carrivick, P.J. (1974). Psychometric, personality and anthropometric variation in a group of Oxfordshire villages. *Annals of Human Biology*, **1**, 365–81.

Herrnstein, R.J. (1971). I.Q in the meritocracy. Allen Lane, London.

Hicks D. (1976). *Primary Health Care*. HMSO, London.

Hooton, E.A. (1939). *The American criminal*. Harvard University Press, Cambridge, MA.

Jones, I.G. and Cameron, D. (1986). Social class analysis—an embarrassment to epidemiology. *Community Medicine*, **6**, 38–48.

Kaplan, B.A. and Mascie-Taylor, C.G.N. (1985). Biosocial factors in the epidemiology of childhood asthma in a British national sample. *Journal of Epidemiology and Community Health*, **39**, 152–6.

Kopec, A.C. (1970). *The distribution of the blood groups in the United Kingdom*. Oxford University Press.

Lasker G.W. and Mascie-Taylor, C.G.N. (1988). Effects of social class differences and social mobility in growth in height, weight and bold mass index in a British cohort. *Annals of Human Biology*, **16**, 1–8.

Leahy, A.M. (1935). Nature–nurture and intelligence. *Genetic Psychological Monographs*, **17**, 235–558.

Logan, W.P.D. and Cushion, A.A. (1974). *Morbidity statistics from general practice, second national study 1970–1*. Studies on Medical and Population Subjects No. 26. HMSO, OPCS, London .

Low, W.D., Chan, S.T., Chang, K.S.F., and Lee, M.M.C. (1964). Skeletal maturation of Southern Chinese children in Hong Kong. *Child Development*, **35**, 1313–25.

McManus, I.C. and Mascie-Taylor, C.G.N. (1983). Biosocial correlates of cognitive ability. *Journal of Biosocial Science*, **15**, 289–306.

MacCarthy, M.D., Douglas, J.W.B., and Mogford, C. (1952). Circumcision in a national sample of 4-year-old children. *British Medical Journal*, **2**, 755–7.

Mascie-Taylor, C.G.N. (1977). Ph. D. thesis, University of Cambridge.

Mascie-Taylor, C.G.N. (1983). Biosocial correlates of IQ. In *The biology of human intelligence* (ed. C.A. Turner and H.B. Miles). The Eugenics Society, London, pp. 99–127.

Mascie-Taylor, C.G.N. (1987). Assortative mating in a contemporary British population. *Annals of Human Biology*, **14**, 59–68.

Mascie-Taylor, C.G.N. (1988). Assortative mating for psychometric characters. In *Human Mating Patterns* (ed. C.G.N. Mascie-Taylor and A.J. Boyce). Cambridge University Press, pp. 201–34.

Mascie-Taylor, C.G.N. and Boldsen, J.L. (1986). Recalled age of menarch in Britain. *Annals of Human Biology*, **13**, 253–7.

Mascie-Taylor, C.G.N. and Gibson, J.B. (1978). Social mobility and IQ components. *Journal of Biosocial Science*, **10**, 263–86.

Mascie-Taylor, C.G.N. and McManus, I.C. (1984). Blood group and socio-economic class. *Nature*, **309**, 395–6.

Mascie-Taylor, C.G.N. and Vandenberg, S.G. (1988). Assortative mating for IQ and personality due to propinquity and personal preference. *Behaviour Genetics*, **18**, 17–24.

Mascie-Taylor, C.G.N., Harrison, G.A., Hiorns, R.W., and Gibson, J.B. (1987). Husband–wife similarities in different components of the WAIS IQ test. *Journal of Biosocial Science*, **19**, 149–56.

Mechanic, D. and Newton, M. (1965). Some problems in the analysis of morbidity data. *Journal of Chronic Disease*, **18**, 569–73.

Nestel, P.D. and Mascie-Taylor, C.G.N. (1987). Nutritional status in Northern Sudan, USAID monographs.

Roberts, D.F. (1977). Assortative mating in man: husband and wife correlations in physical characteristics. *Bulletin of the Eugenics Society*, supplement **2**, 1–45.

Roberts, D.F., Wood, W., and Chinn, S. (1986). Menarcheal age in Cumbria. *Annals of Human Biology*, **13**, 161–70.

Rose, S., Kamin, L.J., and Lewontin, R.C. (1984). *Not in our genes*. Penguin Books, London.

Rutter, M.L. and Madge, M. (1976). *Cycles of disadvantage*. Heinemann Educational Books, London.

Skodak, M. and Skeels, H.M. (1949). A final follow-up study of 100 adopted children. *Journal of Genetic Psychology*, **75**, 85–125.

Smith, A. (1987) Social factors and disease; the medical perspective. *British Medical Journal*, **24**, 881–3.

Terrell, T.R. and Mascie-Taylor, C.G.N. (1990). Biosocial correlates of stature. *Journal of Biosocial Science* (in press).

Waller, J.H. (1971). Achievement and social mobility: relationships among IQ score, education and occupation in two generations. *Social Biology*, **18**, 252–9.

Index